Anonymous

Exposures of Quackery

Being a Series of Articles upon and Analysis of Various Patent Medicines. Vol. 1

Anonymous

Exposures of Quackery
Being a Series of Articles upon and Analysis of Various Patent Medicines. Vol. 1

ISBN/EAN: 9783337779030

Printed in Europe, USA, Canada, Australia, Japan

Cover: Foto ©berggeist007 / pixelio.de

More available books at **www.hansebooks.com**

EXPOSURES OF QUACKERY.

VOL. I.

"TO QUACK OF UNIVERSAL CURES."—HUDIBRAS.

EXPOSURES OF QUACKERY:

BEING A SERIES OF ARTICLES UPON, AND ANALYSES OF,

VARIOUS PATENT MEDICINES.

BY THE
Editor of "HEALTH NEWS."

VOLUME I.

LONDON:
THE SAVOY PRESS, LTD., SAVOY HOUSE, 115, STRAND, W.C.

PRICE ONE SHILLING.

'HEALTH NEWS'
(ILLUSTRATED),
Monthly, 3d.   Post Free for 4 Stamps.

The Best, Cheapest & Most Widely-Circulated Health Journal.   Eleventh Year of Publication.

The Official Organ of the Anti-Adulteration Association.

The Number for January, 1897, begins a New Volume, but subscriptions can commence from any date. Post free for 12 months, 4s.

Devoted to the consideration of Public and Individual Hygiene, House Construction, Dietetics, Foods, Beverages, Adulterations, Health Resorts and Mineral Springs, Domestic Sanitation and Regimen for Invalids, Sanitary Inventions, Literature, &c.

Excellent Advertising Medium for all announcements intended to reach the well-to-do Classes. Rates Moderate. Quack advertisements are rigorously excluded.

THE SAVOY PRESS, Ltd., SAVOY HOUSE, 115, STRAND, LONDON.

PRINTING and PUBLISHING.

THE SAVOY PRESS, Ltd.,
SAVOY HOUSE, 115, STRAND, W.C.,
Are prepared to undertake the
PRINTING AND PUBLISHING OF WORKS IN EVERY CLASS OF LITERATURE AT MODERATE CHARGES.
MSS. read and Confidentially advised upon.

ESTIMATES FREE.

CONTENTS OF VOL. I.

                                                  PAGE

PREFACE 7

CHAPTER I.—Patent Medicines; Patent Medicine Law; Mattei's Electricities 13

CHAPTER II.—Clarke's Blood Mixture .. 18

CHAPTER III. — "Protected by Government Stamp"; Chlorodyne, and other Opiates and Anodynes 23

CHAPTER IV.—Revalenta Arabica 29

CHAPTER V.—The History of Patent Medicines; The Sequah "Prairie Flower" Mixture, and Oil 33

CHAPTER VI.—Holloway's Pills and Ointment; Sequah's Prairie Flower; Clarke's Blood Mixture 47

CHAPTER VII.—Saved from the Waste Paper Basket; Correspondence concerning Holloway and Mattei 57

CHAPTER VIII.—Allen's World's Hair Restorer; Mexican Hair Renewer; Singleton's Golden Ointment for the Eyes; Rowland's Kalydor, and Gowland's Lotion for the Skin; Anna Rupert's Skin Tonic 65

CHAPTER IX.—Quack Advertisements and Testimonials; Mother Seigel's Syrup 76

CHAPTER X.—Clarke's Blood Mixture; The Alleged Testimonial from the late Dr. Swaine Taylor, F.R.S.; The Obverse and the Reverse 88

CHAPTER XI.—Quack Testimonials; Mother Seigel; Clarke's Blood Mixture 94

CHAPTER XII.—Beecham's Pills 97

CHAPTER XIII.—The Alofas (All-a-Farce) Safe Remedies, and the School of Safe Medicine.. 105

CHAPTER XIV.—The Ignorance of Quacks; The Blindness of their Dupes 112

CHAPTER XV.—The Gold Cure for Drunkenness 116

PREFACE

"MAN is a dupable animal," wrote Southey; "quacks in medicine know this, and act upon that knowledge." An older writer, Sir Thomas Browne, M.D., as able a physician as he was a sound philosopher, keenly observed: "Men often swallow falsities for truths," a fact which is even more evident now than it was in his days.

It is true that the old wandering mountebanks and peripatetic charlatans, who frequented fairs, wakes, and such-like gatherings, where empty pates and full purses generally abounded, have almost ceased to exist, except when some travelling representative of Sequah and Co., Limited, drives his gaudy waggon into a country market place, and vends his common aloes—otherwise "Prairie Flower" mixture—and his fish—oil and turpentine—*alias* Sequah's Oil—to gaping rustics, while he now and then draws their teeth and cash to the deafening music of a brass band. "The old order is changed," but the new—what is it? A set of men striving to get rich, working upon the ignorance, the weakness, often the helplessness, of those whom they make their dupes; either telling the most unblushing lies, or carefully dodging the truth in newspaper advertisements and other puffs, the better to create a brisk sale for their rubbish; caring nothing for the health of their customers, but perpetually striving to induce them to " swallow falsities for truths"; and, moreover, availing themselves of the misleading Patent Medicine Stamp, the payment of which comes practically from their dupes, not from themselves, to give colour to the veiled insinuation, or the bold assertion, that their potions, pills, plasters, and other proprietary articles are so valuable that they are " protected by the Government stamp."

As is shown in the following pages, one outcome of this tax is that it leads many thousands of persons to put a fictitious estimate on these bogus remedies. During the past year the revenue derived from patent medicine stamps amounted to £240,062, showing an increase of £14,361 over the sum derived from a similar source in the previous twelve months. Licenses for the sale of these articles have also increased by 1,340 in England, and 111 in Scotland, from which the revenue obtained by Government was £7,188. These licenses are granted, it should be borne in mind, to grocers, drapers, and general stores—in fact, to every applicant willing to pay a few shillings annually for the privilege, without the least regard being had to the fact that many of the licensees have no knowledge whatever of the often dangerous drugs which they sell and recommend to their customers.

The most elementary acquaintance with the action of certain drugs must convince any reflecting individual of the wholesale public injury thus done. To advise, indiscriminately, the use of such powerful medicinal agents as opium, prussic acid, iodide of potassium, aloes, and nitre, for instance, cannot fail to do frequent and great harm. Yet this is what is done every day in our towns and villages throughout the whole length and breadth of the Kingdom. The sole object in taking out a license to sell patent medicines is to increase the licensee's income; and in his ignorance, or through his natural desire to obtain custom, the licensee will lose no opportunity that may present itself for getting rid of his wares, by recommending opiates of variable strength for young children; iodide of potassium (Clarke's Blood Mixture) for anæmic persons; aloes (Mother Seigel's Syrup, Sequah's Prairie Flower Mixture, Beecham's Pills, Holloway's Pills, Williams' Pink Pills, &c.), to delicate women; nitre (Warner's Safe Cure) to individuals suffering from kidney disease, and so on through the list.

All this public injury results from successive Chancellors not caring to disturb a source of yearly income, under a quarter of a million pounds sterling.

Three obvious remedies for this unsatisfactory state of things will suggest themselves, namely: 1, Requiring, as is already done in many other countries, that every proprietary medicine shall bear upon each bottle, box, or packet in which it is sold a conspicuous label stating its composition; 2, withholding licenses to sell patent medicines from all persons having no knowledge of the real nature and action of what they are selling; or, 3, as the question of the sale of patent medicines is considered by the majority of our legislators only as a financial one, very largely increasing the stamp duty on patent medicines, as well as the amount charged for licenses.

That the English are much behind several other nations, as regards the enactments regulating sale of patent medicines, is evident from the following facts:

In France the holder of a pharmacy is forbidden, under very heavy penalties, to sell secret remedies, or even to keep them on his premises.

In Germany the chemist may sell proprietary medicines when ordered by the prescription of a physician; but he must not sell secret remedies. All proprietary medicines sold by the chemist must be prepared under special supervision, and according to the rules of the national pharmacopœia.

In Italy the following regulations as to proprietary medicines came into force on January 1st, 1891:— The composition, as to the quality and quantity of the active substances contained, must be printed on the labels and advertisements; no special curative virtue shall be attributed to them either on the labels or in the advertisements; they shall be sold only by chemists under the surveillance of the sanitary authorities, and with medical prescriptions.

In Japan a public laboratory has been established for the analysis of chemicals and patent medicines. The proprietors are bound to supply a sample of each, with the names and proportions of the ingredients, directions for its use, and an explanation of its alleged efficacy.

In the articles reprinted in the following pages from the HEALTH NEWS in which they appeared, we have endeavoured to show things in their true light. We have often had occasion to speak in plain language, and to call a spade by its proper name; but as Shakespeare says in *King Henry VI.*, " We'll maintain our words," if need arises. We have not extenuated, nor, on the other hand, have we set down aught in malice. We have had in view only the public good, and, judging by the many favourable Press notices, and the still more numerous letters which have reached us from subscribers and other correspondents living in every civilised country in the world, we believe that our efforts have been both appreciated and productive of benefit. The kind encouragement which we have received will stimulate us to fresh attacks upon the hydra-headed monster, Quackery, in future numbers of our journal.

THE EDITOR.

HEALTH NEWS Office:
Savoy House, 115, Strand, W.C.

Exposures of Quackery.

CONTENTS OF VOL. II.

CHAP. 1. "Pink Pills for Pale People."—CHAP. 2. Warner's Safe Cure.—CHAP. 3. Quack Advertising; Clarke's Blood Mixture and the False Testimonial.—CHAP. 4. Anonymous Abuse; Warner's Safe Cure and Medical Staff; A Quack Libel Case; Morison's Pills; Dixon's, Fothergill's, and Lee's Pill's.—CHAP. 5. Electric Belts; Mattei's Electricities; Nicholson's Ear Drums.—CHAP. 6. St. Jacob's Oil; Mother Seigel's Syrup.—CHAP. 7. Our Correspondents and Critics; Silverton's Remedies for Deafness; Unqualified Practitioners; "A Merciful Medicine, more Precious than Rubies."—CHAP. 8. Patent Medicines and Pious Language; The "Reverend Specialist"; Congreve's Balsamic Elixir; Owbridge's Lung Tonic; Lane's Catarrh Cure; A Quack's Certificate.—CHAP. 9. Stepney Green Diplomas.—CHAP. 10. Bone Setting: A Patent Medicine Song.—CHAP. 11. Quack Bogus Newspapers; How the Poor are Swindled; Handyside's Consumption and Cancer Cure; Electric Snuff.—CHAP. 12. Mattei's Electricities in Court; A Curious Way of Exposing Quackery.—CHAP. 13. Patent Medicine Testimonials; St. Jacob's Oil; Clarke's Blood Mixture; The Man who gave himself a Testimonial; Eno's Fruit Salt.—CHAP. 14. "Handy" Still Shuffling.—CHAP. 15. Quack Newspapers; The Sequah Bubble Burst.—CHAP. 16. Fenning's Fever Curer.

PRICE ONE SHILLING,
Post Free for 14 *Stamps to any part of the Kingdom.*

THE SAVOY PRESS, LTD., Savoy House, 115, Strand, W.C.

CHAPTER I.

PATENT MEDICINES; PATENT MEDICINE LAW; MATTEI'S "ELECTRICITIES."

For the sake of realising an annual sum of money—about £250,000—the British Government perpetrates a gross wrong on the community by giving a fictitious value and importance to the myriad quack remedies which are sold throughout the country. We refer, of course, to the amount raised by issuing Government Stamps for articles of this class, designated as Patent Medicines; but, as we are in the habit of calling a spade a spade, we prefer to use the shorter name. The injury done to the community by this system of protecting quack medicines is incalculable. "Oh!" say the artisan, the small tradesman's wife, and many others in a higher position in society, "this must be a good remedy, it is protected by Royal Letters Patent." And so they go on, swallowing the bait as well as the physic, whilst the proprietor reaps a golden harvest. Government officials are always slow in making any changes involving them in additional trouble or new arrangements; and, consequently, succeeding Chancellors of the Exchequer leave the question of the Patent Medicine Stamp alone. Of late years there has been—and very properly, too—considerable public agitation concerning the adulteration of articles of food and drink, and it has been enacted that various admixtures of this character can only be sold with a distinctive label showing their composition. But there is no provision of this nature on behalf of the purchasers of quack medicines, for which the term "Patent" is a misnomer, and also misleading, as the only protection afforded by the Government stamp is to the manufacturer of the nostrum, all information concerning the ingredients that enter into its composition being withheld.

Patents for new and useful inventions carry with them the sole right of making, using, or selling such inventions for a limited period only, namely fourteen years, and before they can be obtained the person seeking the patent must deposit at the Patent Office a description of the invention. Moreover, he has to make large and frequent payments, otherwise the patent lapses. In the case of "Patent" medicines, nothing is required by way of proving either novelty or utility, while the composition remains a mystery. All that the manufacturer has to do is to pay for a given number of stamps, only three-halfpence each, and he can dose as many people as he can induce to buy the stuff with the most noxious drugs, or he can make a pile of money by selling coloured water, soap, or any other commodity, very cheap when purchased in bulk, and capable of being disguised in appearance or taste before it is passed on to the consumer in smaller quantities.

A controversy was for some time carried on in one of our contemporaries, the *National Review*, concerning the "Electrical Remedies" introduced by an Italian Count. This discussion would certainly have terminated at an earlier stage had it sooner occurred to Dr. Herbert Snow, one of the medical staff of the Cancer Hospital at Brompton, to submit samples of Mattei's "Electrical Remedies" to a competent analyst. He did so, however, and the subjoined report made by the well-known chemical authority, Mr. A. W. Stokes, is worthy of study.

Analytical Laboratory, Vestry Hall,
Paddington Green, W.,
October 21st.

Dear Sir,—On October 2nd I received from you three small bottles bearing the Government patent medicine stamp, each securely sealed with a wax, unbroken, seal of a castle on a rock. I have now carefully examined these chemically, physically, and microscopically, and I find as follows:—

They were labelled, "Elettricita Bianca," "Elettricita Verde," and "Elettricita Rossa."

To find if they possessed any special magnetic properties, they were placed singly in thin glass tubes; these tubes were suspended by silk filaments. Under such circumstances, an electrical body would point one end to the north and the other to the south. Not one of these came to rest in such a position.

neither were any of them attracted by a magnet as a magnetic body should be. Hence they certainly are not magnetic. Other tests showed that they were not electrical.

To delicate test-paper they were perfectly neutral. Vegetable extracts are usually either alkaline or acid; even if neutral when fresh, they speedily change.

They had the following characters:—

	Elettricita Bianca. (White Electricity.)	Elettricita Verde. (Green Electricity.)	Elettricita Rossa. (Red Electricity.)
Colour	None	None	None
Odour	None	None	None
Taste	None	None	None
Polarity	None	None	None
Specific gravity (distilled water=1)	1·0006	1·0002	1·0002
Solid matter in 100 parts	0·01	0·01	0·01
Metals *	None	None	None
Alkaloids	None	None	None

The microscope showed an absence of any floating particles or sediments such as are usually present in vegetable extracts.

There is but one substance which possesses all the above qualities—that is *water*.

None of these fluids differ at all from water in any of their properties.

Yours faithfully,

ALF. W. STOKES, F.C.S., F.I.C.,
Public Analyst to Paddington, Bethnal Green,
and St. Luke's; Gas Examiner to the
London County Council.

---

* By metals is meant any foreign to water, or any such as are used medicinally.

The analysis speak fully for itself. The Count has "beaten the record" in producing an article absolutely negative in its character, and there can be no other feeling than one of indignation when we think of the poor creatures who have been deluded into the use of the Count's specifics—white, green, and red. But even in description of colour the Count has broken faith with his credulous customers, for, according to Mr. Stokes, the green and the red electricities are as devoid of colour as the white. Possibly the green may have been so named out of a delicate compliment to the Count's admirer's, while the absence of coloration may be intended to heighten the joke. If the Count has been oblivious as regards the colour, he has not forgotten the charge; for these *precious* electric remedies, which, according to the Count's dupes, will, when administered in drops, cure cataract, mend broken bones, and remedy every other ill to which human nature is liable, are sold at the rate of 3s. 9d. for a small phial containing three-fourths of an ounce. As soon as the money market has recovered from its low, unspeculative condition, we shall really expect to see in the financial papers an advertisement worded as follows:—

"Wanted, a few persons to join a syndicate for purchasing water at 1s. per 1,000 gallons, and retailing it at 5s. an ounce!"

Chapter II.

Clarke's Blood Mixture.

In our first chapter we pointed out that the word "patent," as applied to patent medicines, is most misleading, for the purchaser is left wholly in the dark as to the nature of what he buys. The consequence is, either that he may lay out his money upon a bottle of water, as shown by the analysis of Mattei's "Electrical Remedies," published in Chapter I., or that he may unwittingly dose himself or some member of his family with the most potent drugs, and thus cause very serious injury to health, even if worse consequences should not result.

Judging by the numerous communications which have reached us, it is evident that the subject of patent medicines is one of general interest, and that a very considerable number of thinking people are of accord with us as to the urgent desirability of amending the law relating to them.

Many of our correspondents ask for information concerning various advertised nostrums. The field is so vast that we are at a loss to decide what preparation to take next—as a topic of discussion only, thank goodness! Well, we will deal with the first letter that comes to hand from a heap in front of us. It is from a lady subscriber who inquires, apparently with no small misgivings (which we shall presently prove to be well founded), whether she, being in delicate health, is right in continuing the use of Clarke's Blood Mixture.

Recollecting that this identical mixture was reported upon by Dr. Alfred Swaine Taylor, F.R.S., years ago, we referred to the back volumes of the *Lancet*, and in that for 1875 we found, forming part of a letter headed "Quack Medicines," a copy of the "Report of Analysis of a liquid described as 'Clarke's World-Famed Blood Mixture or Purifier,'" by Dr. Taylor, the late eminent analyst and lecturer on medical jurisprudence at Guy's Hospital. The examination of an eight-ounce bottle of the mixture showed the ingredients to be as follows:—Iodide of potassium, 64 grains; chloric ether, 4 drachms; solution of potash, 30 minims; water coloured with burnt sugar to give the requisite tint, $7\frac{1}{2}$ ounces. The dose directed to be administered was one tablespoonful (half-an-ounce) four times a day. "Why such a mixture as this," says Dr. Taylor, "should be designated a 'blood mixture' and a 'blood purifier' is incomprehensible. It has no more claim to this title than nitre, common salt, sal ammoniac, or other saline medicines which operate on and through the blood by absorption. Its properties (*i.e.*, those of iodide of potassium) are well known, and there is no novelty in its employment. The only novelty in this form of mixture is that the iodide is dissolved in water coloured with burnt sugar, and that it is described as a 'blood purifier.' The four doses directed to be taken daily represent sixteen grains, and if the person taking it is not under medical observation, such a daily quantity as this may accumulate in the system and do mischief. In some constitutions the iodide of potassium frequently taken proves specially injurious. It produces iodism." Iodism is the condition when symptoms of special poisoning show themselves, similar to the salivation, &c., in cases of mercurial poisoning.

We have purposely quoted at some length from Dr. Taylor's report, partly because it emanates from such a high and unimpeachable authority, partly because the facts are stated by him with marked moderation.

As regards the deleterious effects of iodide of potassium, in unsuitable cases, or in long-continued doses, all other medical authorities fully agree with Dr. Taylor. For instance, Dr. Sidney Ringer, physician to and lecturer at University College Hospital, writing about this drug in his "Handbook of Therapeutics," says that if its administration is continued for a long period, or if the patient manifests great susceptibility to its action, iodism is produced; also that this condition may arise after very small doses The parts chiefly affected in iodism are the eyes, the nose, the mouth, the stomach, and the bowels; there is also sometimes a distinct skin eruption. Inflammation of the mucous membranes covering the eyes, running at the nose, a form of salivation resembling that caused by mercury, purging, and nausea, with loss of appetite, all or some of these symptoms will then make their appearance. "A grain or even less," writes Dr. Ringer, "may affect the stomach"; moreover, he observes elsewhere, "Iodide of potassium sometimes produces distressing depression of mind and body. The patient becomes irritable, dejected, listless, and wretched. Exercise soon produces fatigue, and perhaps fainting."

We could quote hundreds of similar proofs of the danger arising from the indiscriminate administration of this powerful drug which constitute the basis of this so-called "blood mixture"; but, surely, enough has been said on this point to convince the most sceptical that iodide of potassium should never be given except in selected cases, and under the supervision of a qualified practitioner.

Yet the printed directions accompanying a bottle of Clarke's Blood Mixture which we recently purchased, after recommending this preparation as a never-failing cure for a whole host of diseases, state that it is "warranted free from anything injurious to the most delicate constitution of either sex."

We will leave our readers to form their own opinion on this question. Before quitting the subject we may mention with regard to the manufacturer of this blood mixture that the analysis of the bottle just referred to, made in the laboratory of Mr. Alfred W. Stokes, F.C.S., public analyst to Paddington, Bethnal Green, and St. Luke's, gives only 48 grains of iodide of potassium in the eight ounces. We infer, therefore, that at some period subsequent to 1875, the manufacturers, out of deference to Dr. Taylor's views, diminished the quantity of this drug so as to reduce the dose from four to three grains. It is a pity that the reduction was not carried still further. Indeed, if the iodide of potassium had been omitted altogether, the warranty might have been given on better ground than now, while a corresponding increase of the water and burnt sugar could have done no possible harm to "the most delicate constitution of either sex."

Chapter III.

"Protected by Government Stamp"; Chlorodyne, and other Opiates and Anodynes.

Amongst the correspondence which these articles have elicited we have received the following letter from a gentleman at Shrewsbury:—"I should like to ask you a question about quack medicines, which perhaps you will kindly answer in an early number of your journal. Of course I do not dispute what Dr. Taylor has said as to the probable effects of taking Clarke's Blood Mixture; but what do you say to the case—which I cannot now find—mentioned in Clarke's advertisements, of a pauper who cured several holes in his legs by taking two or three bottles of the blood mixture? Did it never happen, or if it did, how was it that the doctors could not cure him?"

Like our correspondent, we cannot find any particulars of the case upon which he seems to think that the reputation of this "world-famed" specific for all the ills that trouble humanity must stand or fall. Advertisements occasionally furnish much amusing reading, but we must confess that when we come to a quack advertisement we commonly pass it by. It may be that we get tired of the monotony of successes all along the line, or that we are too matter-of-fact in our views to believe in modern miracles; consequently we have not made, even in print, the acquaintance of the pauper referred to. Our correspondent will therefore, we trust, see our inability to discuss upon its merits a case of which we possess no knowledge whatever. Besides, our correspondent is somewhat wide of the mark. Our object in writing this series of articles is to urge the necessity of altering the present system of issuing Government stamps for quack preparations, seeing that a very large proportion of the community entertains the mistaken notion that the words "Protected by Government Stamp" convey some sort of guarantee as to the value of the ingredients. Unfortunately, the reverse of this is the fact; and hundreds of thousands of gallons of quack medicines, containing drugs potent for harm, are under the ægis of the Government stamp distributed throughout the length and breadth of the land. Further, we showed in the previous chapter that the basis of Clarke's Blood Mixture is iodide of potassium, a drug of such powerful action that, in many persons, even when administered in very moderate doses, it produces most distressing symptoms, while it should never be given for any length of time except under proper medical advice and supervision. When this general statement has been refuted, we shall be quite ready to deal with individual cases alleged to have been cured. If the iodide of potassium were as innocuous in its nature as the burnt sugar with which the Blood Mixture is flavoured and coloured, there would be admittedly less reason for these strong remarks made by Professor A. Swaine Taylor, F.R.S.: "The sale of medicines of this kind should be strictly prohibited, unless the bottles containing them were issued with a caution label setting forth their true composition. It is only reasonable that a person should know what he is purchasing." Various Continental countries have recognised this necessity, and have framed laws prohibiting the sale of any alleged remedies, of which the composition is not publicly made known.

According to the Sale of Poisons Act, it is specially laid down that certain regulations must be observed with respect to the selling of different poisons, the penalty for the breach of these regulations being £5 for the first offence, and £10 for each subsequent offence. It is enacted that upon the sale of any of these poisons, the box, vessel, or cover in which it is contained must be labelled with the name of the article, the word "poison," and the name and address of the vendor. With respect to poisonous vegetable alkaloids and their salts (hydrochlorate of morphia, for example), no such article may be sold to any person unknown to the seller, unless introduced by some person known to the seller, and upon every such sale the seller must, before delivery, enter in a book to be kept for that purpose the date, the name, and address of the purchaser, the name and quantity of poison sold, and the purpose for which it is required, and must also cause the purchaser and the person introducing the purchaser to sign their names therein. Yet the provisions of this important Act are openly violated by the sale of preparations containing these poisons and bearing the Government patent stamp. Take the case of chlorodyne, for instance. The analysis of the widest-known preparation bearing this name, sold as Collis Browne's Chlorodyne, was made for us by Mr. A. W. Stokes, F.C.S., F.I.C., and that gentleman reported that in an ounce bottle he found six drachms of chloroform, a small quantity of Indian hemp, and six grains of hydrochlorate of morphia, with some unimportant ingredients. An analysis, published some years ago, by Dr. Wynter Blyth, gave, in addition to the various poisons enumerated by Mr. Stokes, twelve drops of Scheele's prussic acid in a rather larger quantity of chlorodyne. Now, supposing that the prussic acid is at the present time omitted, what conclusion can we come to as regards the two poisons in largest proportion, namely, chloroform and hydrochlorate of morphia, both of which are of course included in the schedule of poisons coming under the Act to which we have referred ?

The dose of hydrochlorate of morphia, when administered medically, is one-eighth of a grain to half a grain; of chloroform, three to ten drops. In one ounce of Collis Browne's Chlorodyne, by Mr. Stokes' showing, there would be the equivalent of twelve full doses of morphia, and thirty-six full doses of chloroform.

If a person wanted to buy six drachms of chloroform, he would most certainly find considerable difficulty in procuring it; but put that amount in a bottle with other poisons, clap on a Government three-halfpenny stamp, and it can be bought with almost as much ease as the harmless treacle with which it is flavoured!

The quantity of this and other chlorodynes sold is something enormous, far surpassing the imagination of anyone who has not given attention to the matter. Taken, at first, in small doses by the unhappy persons who drug themselves with chlorodyne, the victims become gradually habituated to its use, and many fall sooner or later, a prey to the craving for morphia. This craving once established, they become as completely slaves to the practice of swallowing chlorodyne in extraordinary quantities as a Chinaman does to smoking opium, or a Malay to using Indian hemp, with the inevitable consequence that physical, mental, and moral deterioration must follow.

A druggist in business at the West End recently mentioned to a friend of ours numerous instances of morphia-craving which had come under his own cognisance. One young woman purchased from him in a fortnight forty-two ounces of chlorodyne, the whole of which quantity she herself consumed. In another case, a man regularly bought every day for years a 4s. 6d. bottle. In a third, a lady consumer ran up a bill for chlorodyne amounting to £13 in six months, in addition to what she paid for at the time of purchase. But even this last case seems moderate in comparison with that of the wife of a well-known actor, who in six months became indebted to her chemist for £90 for chlorodyne, a matter which the circumstance of the bill being disputed brought into public knowledge.

The revelations made at the inquest on Dr. Lyddon, in what was styled the Faversham Mystery, showed how feeble in body and mind the votaries of morphia-drugging become; while, in a *cause célèbre* in Paris, it was brought out in evidence that the unfortunate lady concerned had expended more than 25,000 francs (£1,000) upon the purchase of morphia, with the most baneful results to her physical and mental condition.

Chapter IV.

Revalenta Arabica.

We saw some time since a monster advertisement, which took up the greater part of a page in a London daily paper, setting forth the marvellous curative virtues of Du Barry's Revalenta Arabica. The advertisement professes to give a great deal of information, though not a word is said as to who Du Barry is. Originally, there may have been a man who slipped "Du" in front of his name because it looked like "Dr.," but at present we understand that "the show" is run by a syndicate.

From asthma right away through the alphabet down to vertigo, no disease has ever been known to resist this wonderful remedial agent, according to the published specimens of the 100,000 (we hope our compositors will be careful with the noughts—there are plenty of them) testimonials of cures which the proprietors profess to have in their possession. Besides, these testimonials are many of them quite respectable by reason of their age, bearing such dates, for instance, as 1850, 1852, and the like, while some may be even older, for they have no date at all.

And such people of rank, too!—for example, a

Marchioness de Bréhan, of Versailles, whose cure is numbered 58,614, felt so " dreadfully low-spirited " that even the voice of her maid annoyed her. There is no novelty in ladies being annoyed at the voices of their maids, particularly if the latter, irritated by their mistresses' scolding, should " give it them back again," as would be said when subsequently describing the incident in the servants' hall. However, the poor marchioness must have been in a very bad way, for she asserts that " many medical men, English as well as French, had prescribed in vain." The last three words remind us of the doggerel epitaph, " Affliction sore long time I bore, physicians were in vain," &c. It rather puzzles us, as the marchioness has not thrown any light upon the matter, to make out how she contrived to consult so many English physicians at the French town of Versailles, but that is a mere detail, which one loses sight of in rejoicing that she not only recovered her health, but was able to resume her social position.

How nice, and how thoughtful of Du Barry & Co. to give us the cases of such interesting people, instead of vulgar, common paupers like the one who is said to have cured the holes in his legs by taking Clarke's Blood Mixture! Looking at the matter from a business point of view, too, it costs no more to advertise a marchioness's testimonial than a pauper's. So our advice to quacks is: Stick to the people of title, and let the paupers go to the—workhouse.

"A fool and his money are soon parted," as our readers will presently see—not that we consider ourselves the fool, having a special reason for the purchase. We invested two shillings in a half-pound packet of Du Barry's Revalenta Arabica at a chemist's shop in the Strand, being particular about its being Du Barry's, as the manufacturers of this article have issued a caution against *cheap* foods. Two shillings for a half-pound cannot be said to be cheap, when we state that it must have cost the manufacturers probably a penny, as we will show directly.

On the next day we forwarded our purchase to an eminent analyst. We will let him take up the story for a while, for we must confess that the following report nearly carried away our breath:—

Analytical Laboratory, Vestry Hall,
Paddington Green, W.,
February 14th.

Dear Sir,—On February 10th I received from you a half-pound tin of "Du Barry's Revalenta Arabica Food." This was enclosed in unopened wrappers. I have now made a careful chemical and microscopical examination of the material.

I am of opinion that it consists solely of lentils ground up into a fine powder. I could detect no added ingredient possessing any medicinal or other properties.

I remain, yours faithfully,
ALF. W. STOKES, F.C.S., F.I.C.,
Public Analyst.

To the Editor.

With this report Mr. Stokes returned to us the wrappers and copy of testimonials which they had contained. The testimonials were, as far as we took the trouble to ascertain, similar to those which attracted our attention in the newspaper advertisement; and there was the same lengthy list of diseases for which, to use Du Barry & Co.'s own words, "this delicious food is the *only* cure." We have put the word "*only*" in italics, because we never before heard of lentils being the only cure for consumption, deafness, diabetes, dropsy, paralysis, and many other serious disorders.

On the wrapper is given a small woodcut of a number of black men, very scantily clad (they look as if Revalenta had almost cured them of clothing), presumably engaged in cultivating lentils, with the words printed underneath:—" Discovered, exclusively grown, and imported by Du Barry & Co."

If Du Barry & Co. really discovered lentils, their firm must, indeed, be of long standing, seeing that this leguminous plant was well known to the Hebrews and other ancient nations. As to Du Barry & Co.'s exclusively growing lentils, we should like to learn where they accomplished this extraordinary feat, and why lentils can be bought at any corn chandler's shop for about twopence per pound, at which rate "lentils ground up into a fine powder" (*vide* Mr. Stokes' report) can be purchased in bulk at Mark Lane.

"In the name of the Prophet, figs," was the cry of the itinerant fruit-seller in the old Eastern tale. "In the name of the *profit*, lentils," is the new reading suggested by Du Barry's Revalenta Arabica.

Chapter V.

The History of Patent Medicines: The Sequah "Prairie Flower" Mixture, and Oil.

One of our colonial subscribers asks us how, having regard to the origin of the word "patent," that term came to be applied to preparations of which the composition is kept secret. The misnomer arose in this way.

During the reign of George III., whose obstinacy and incompetence lost our fairest colony, America, and raised the National Debt from £138,000,000 to £794,000,000, the Ministers were frequently much exercised in their minds as to the means of raising money. The Chancellor of the Exchequer of that period bethought him of quack medicines, and an Act of Parliament was passed (23rd George III., cap. 62) "to grant to His Majesty a stamp duty on licenses to be taken out by certain persons uttering and vending medicines, and certain stamp duties on medicines sold under such licenses, or under authority of His Majesty's Letters Patent." A subsequent Act, in 1785, designated these medicines as "prepared or compounded by any person or persons whatsoever" who had or claimed to have "any occult or secret or unknown art, or some exclusive right or title to their manufacture," the same being advertised or recommended as "specifics or otherwise for the cure or relief of any ailment or malady incident to, or in any way affecting, the human body." The *otherwise* in this definition applies quite as much to quack medicines at the present day as it did then.

It will be seen that the term "patent" applied to these secret quack compounds is used in the sense of "privileged" (by letters patent), and not of "open or divulged."

In 1875 an act was passed reducing the medicine license duty, which previously had ranged from five shillings to forty shillings in different localities, to the uniform amount of five shillings throughout the United Kingdom. The consequence of this reduction was to greatly increase the number of patent medicine vendors. In 1874, the year before the passing of the new Act, the licenses taken out were 12,430; now the annual number is about double that—considerably over 20,000, The question naturally arises, Who are these patent medicine vendors? The official register of chemists and druggists for a recent year, kept by the Pharmaceutical Society of Great Britain, contained 13,812 names. Deducting one-third, as being assistants and not carrying on business on their own behalf, we can account for (roughly estimating) 9,000 chemists, leaving nearly 15,000 licenses in the hands of various tradesmen—grocers, drapers, and general shopkeepers—wholly unacquainted with even the rudiments of chemical knowledge, yet authorised by law to deal in secret medicines, often of the most dangerous character. The ease with which these can be procured—we refer now more particularly to the opiate and other narcotic preparations—has led to a widespread system of home-drugging, while it is undoubtedly responsible for many deaths, especially of children, not invariably arising from mere misadventure, but always very difficult to detect and bring to light.

Well might a coroner of large experience remark that "it is impossible to say how many infants are killed annually by soothing syrups." Every preparation of this class that we have examined contains opium in some form or other.

Various remedies have been suggested for this most unsatisfactory condition of things. Some propose the increase of the now very small license duty to such an amount as would deter many of the present holders from dealing in such questionable and even dangerous merchandise. Others, again, urge the repeal of the stamp duty on patent medicines, because it conveys to the public a sort of Government guarantee. To our mind the question of the increase or removal of the stamp duty is of a comparatively minor character, and, as it is of a fiscal nature, might be left to the consideration of the Chancellor of the Exchequer for the time being. We should, however, lean to an increase of the stamp duty, and to a very great increase of the charge for the patent medicine license, which ought to be restricted to chemists and druggists alone, so as to facilitate bringing the articles sold as patent medicines under the provisions of the Pharmacy Act, 1868, also called "An Act to regulate the sale of poisons"; out of which section xvi., enacting "that nothing hereinbefore contained shall extend to or interfere with the making or dealing in patent medicines," should be removed.

But the great reform necessary is to take a leaf out of the legislature of Continental countries, and to make it unlawful to sell any patent medicine without previously placing on the bottle or box in which it is issued to the public a legible description of the contents. By law the sale of certain articles of food, such as chicory and margarine, is prohibited without a label apprising the purchaser of the nature of the substance which he is buying. How much the more is it desirable, in the interests of the purchasers of patent medicines, that they should be duly warned as to the composition of articles which contain potent drugs inimical to health and often dangerous to life?

Further, if a stamp duty be continued, it should be clearly made known that it is to be regarded as of the nature of a tax, and not—as now commonly supposed—as an authoritative guarantee of the quality or value of the preparation. "Protected by Her Majesty's

Royal Letters Patent" is a statement which has deluded hundreds of thousands of purchasers of quack remedies.

But if the manufacturers of quack medicines are silent concerning the actual ingredients of the stuff concocted by them, they cannot be charged with similar reticence respecting the pretended virtues of such components. Bold assertion is their sheet-anchor, and equally bold advertisement is the talisman with which they conjure the coin out of the pockets of their credulous customers. The proprietors of quack medicines are evidently indoctrinated with the views expressed by Carlyle when he said, "Great Britain contains so many millions of people—mostly fools," and they go for the "mostly" with an energy and determination worthy of a better cause, sparing no expense, and stopping short at no assurances.

Some little time back we bought a small bottle (two ounces for two shillings) of Sequah's "Prairie Flower," and sent it to an analyst for the purpose of examination. But why examine it (some may ask) in face of the positive statements and certificate published in the prospectus which accompanied our purchase? Well, we like to be independent, and as we did not stand in need of Sequah's stuff ourselves, while we have too much regard for our quadruped friends to experiment on them, we had no alternative between smashing our shilling per ounce treasure or forwarding it to Mr. Stokes, so to his laboratory it went

In a circular headed "Sequah's 'Prairie Flower' and all about it" is a most glowing description of the mineral springs of the Pacific slope of North America. From this the transition is easy to the wonderful springs on the borders of the Montana Territory, "the most noted of them all" having been secured by purchase by Sequah, Limited, "as far as Europe and the United Kingdom are concerned." If anyone should doubt this, he can test it by application to the "Company's London solicitor," in whose hands the legal documents have been placed; but, through some singular oversight, both the name and address of this official have been omitted. Next, we learn that the Edinburgh *Evening Dispatch* (no year given) suggests that invalids, to obtain benefit, must go out and take the waters on the spot. "This is all very well," continues the compiler of the Sequah circular, "so far as a few wealthy individuals are concerned, but it is utterly impossible for the poor, for business men, and for the bulk of the middle classes, who can spend neither the time nor the money for so costly a trip. But as the people cannot go to the springs," the writer adds (displaying his ecstasy in capital letters), "SEQUAH HAS BROUGHT THE SPRINGS TO THE PEOPLE. It was found that the water could be concentrated by careful evaporation and still retain its curative virtues; and SEQUAH, LIMITED, took advantage of this fact, and made it their business to bring these waters before the public in a CHEAP AND CONVENIENT FORM. For Rheumatism is closely associated with the great group of STOMACH AND LIVER COMPLAINTS, and it was found necessary, in order to ensure a COMPLETE AND PERMANENT CURE, to combine this mineral water with certain VEGETABLE EXTRACTS, several of which are also valuable Indian medicines, and found in the woods and prairies of the Far West. And thus compounded, PRAIRIE FLOWER is undoubtedly far and away the best remedy ever yet introduced for all sorts of complaints and other CHRONIC DISEASES." Who but a sceptic could doubt the assurances conveyed with the aid of capital letters, ordinary type being quite inadequate for the purpose?

Moreover, we learn from another circular that "this wonderful and world-renowned preparation has been in use amongst the Sioux, Cherokees, Comanches, Apaches, and several other tribes of North American Indians for hundreds of years." Steady, there, Messrs. Sequah, Limited! If the North American Indians have used this "wonderful and world-renowned preparation" for a period extending as far back as the discovery of America, how comes it that you claim to have invented it? We enjoyed for many years a friendship with the late Mr. Catlin, the eminent traveller, who spent much of his life (more than a quarter of a century) amongst the native tribes of North America, and we have, on numerous occasions, discussed with him the primitive remedies employed by the Indians. Never, either in his conversations or in the published accounts of his travels, did Mr. Catlin even refer to the "Prairie Flower." Yet he was a keen observer and a careful recorder of all that related to the habits and domestic customs of the Indians.

The fact is, that the Sequah writer is all wrong, for aloes, the "botanic extract" found in the "Prairie Flower," is obtained from the East and West Indies, where the North Americans Indians are not likely to have gone in search of that drug, so that the "Far West" statement is obviously *far fetched*. There is absolutely nothing new in the administration, medicinally, of such an ancient pharmaceutical acquaintance as aloes. Consequently, the sole novelty consists in falsely describing the countries whence this drug is procured. Another matter to which we would draw attention is that aloes is uncertain, and in the case of very delicate persons and children often injurious, in its action, unless regulated by other medicines combined with it; but if Sequah, Limited, take so little trouble as not to learn whence a drug comes, they can hardly be expected to inform themselves as to its properties.

At the same time that we sent the "Prairie Flower" to our analyst, we also submitted a bottle of Sequah's Oil for examination. The following is a copy of his report upon both articles :—

Analytical Laboratory,
Vestry Hall, Paddington Green, W.
March 13th.

Dear Sir,—On February 20th I received from you a two-ounce bottle of Sequah's Oil and a two-ounce bottle of Sequah's "Prairie Flower." These were still in unopened wrappers, and were sealed by the unbroken patent stamp of the Inland Revenue.

On analysis, I found as follows :—

The "Sequah Oil" consists of a mixture of two-thirds turpentine and one-third fish oil, scented with a few drops of oil of camphor.

The Sequah "Prairie Flower" contains in two ounces :—

    Water ..   735 grains
    Aloes ..   ..   105 ,,
    Carbonate of soda   ..   .   35 ,,

and a few drops of the tinctures of capsicum and myrrh.

This medicine being reputed to be made from "a Mineral Water and Vegetable Extracts found in the Far West," I carefully looked for the usual constituents of ordinary mineral water; but, excepting the carbonate of soda, I found none.

        I remain,
                Yours faithfully,
                  ALFRED W. STOKES, F.C.S., F.I.C.,
To the Editor.   Public Analyst.

The carbonate of soda is doubtless added for the purpose of keeping the aloes in a state of solution, so that it looks very much as if the vaunted "Mineral Water" came from no more remote source than the ordinary house-tap.

As regards the Sequah Oil, that is decidedly fishy, like the method resorted to by Sequah, Limited, for getting publicity for their wares. The bulk of quack preparations obtain a notoriety through newspaper advertisements and mural posters; but Sequah, Limited, taking advantage of the falsely assumed connection between a watery solution of East and West Indian aloes and the North American Indians, send out itinerant lecturers—half sham-Indian, half English " Cheap Jacks "—into the market-places and highways to spout rhapsodical praises of Sequah This and Sequah That. These lecturers, more numerous by far than the knaves in several packs of cards (we mention the knaves through a natural sequence of thought), are distributed in various towns, and succeed at any rate in one thing, and that is in disturbing the peace and quiet of any locality that may be pestered with their presence. Being glib of tongue, ready with clap-trap talk, and liberal in treating the mob, they soon get around them a set of partisans, whose number is rapidly swollen by loungers and others.

In the course of a recent trial for slander, the defendant being a Sequah lecturer at Croydon, it was stated by counsel that the total number of these men was upwards of a dozen; nearer a score, indeed. Yet their hearers are led to foolishly imagine that each is the *only* " Sequah "; while, as a matter of fact, there is not a single fellow amongst the whole crew entitled to the name.

A correspondent has favoured us with a description of one of these meetings, from which we will quote :—

"In a field there was drawn up a highly-decorated circus-car, which during part of the day had paraded the streets with a brass band sitting in it. Flaring lamps fixed on poles served to brilliantly light the scene. Preceded by a band, the Sequah lecturer drove up to the car in a two-horse landau. Taking a seat in the front of the car, he proceeded to produce letters from his pocket, and open and silently peruse them, apparently jotting down their contents. Meanwhile he would now and then be interrupted by messengers bringing bouquets of flowers, fruit, and more letters. After about ten minutes or more of this elaborate dumb show, he would rise and remove his overcoat and broad-brimmed sombrero hat, showing a mass of black, greasy, wavy hair, a string of boars' tusks round his neck, and garments of the cowboy style with which Buffalo Bill's followers made us familiar. Then he would read aloud a number of letters of thanks for cures effected, or asking for more medicine. After a little talk about the extent to which the Sequah remedies were spreading, he fitted on his forehead a small electric lamp, the portable battery of which was carried in his pocket. Those persons who desired to have teeth removed were invited into the car. I saw between thirty and forty men, women and children step up, one at a time, and, with the aid of the electric lamp, their teeth were taken out, some having as many as three or four extracted. The operation was rapidly performed, but an ominous snap now and then told to the initiated that a tooth had been broken off. When the cries of the sufferer who was being operated on were loud and expressive, the brass band stationed at the rear portion of the car struck up lustily, so as to drown the sound, and the only air that they played during the whole time was singularly selected, being the negro melody, "Who's dat calling so sweet?" Then came more boasting and more letter reading. After this, people in the crowd who had been cured of rheumatism were requested to come into the car.

Some four persons accepted this invitation, bent their arms up and down, which they said they had not been able to do previous to treatment, or jumped about before the spectators. Then followed the sale of Sequah preparations at two shillings a small bottle, and the stuff went off at a lively rate. Occasionally people having rheumatic limbs were rubbed in the car by the lecturer or his attendant for half-an-hour or more, and sometimes said they felt better. When the rheumatic cripples had ascended the car, the lecturer generally broke their crutches, then 'massaged' their limbs, and finally bade them to walk away. One unfortunate fellow hobbled from the car to the edge of the field; but, crutchless as he was, he could get no further, despite the vigorous and long rubbing he had been subjected to."

It is now well understood that any benefit which the sufferers may receive is entirely due to the "massage," or careful rubbing; also that the cases are *selected*. There is no inherent curative virtue in the Sequah Oil, as shown by the circumstance that, in the majority of cases of alleged cure, the relief is not permanent, while home trials commonly fail altogether.

Even the so-called cures are not all so genuine as they are made to appear, as the following anecdote will prove. A lady residing at Surbiton Hill had a gardener, who, though somewhat stiff in his joints, after the wont of gardeners through the nature of their employment, could do a good day's work every week-day. When a Sequah lecturer was at Kingston-on-Thames, this lady heard so much about him that she made up her mind to drive down and hear him hold forth. As she sat in her carriage outside the concourse of people, what was her astonishment, on seeing a patient hoisted into the car (with much labour by four men), at recognising in him her gardener! After a while she saw him run nimbly down the steps. Next day she taxed him with all this, and said, "Surely, gardener you could have got into the car without giving all that trouble?" To which he replied, "Oh, yes, Mum, but them's the orders!" From other reports which have reached us, it is evident that whatever doubt there may be as to the diseases which the Sequah remedies can relieve, there is room for none as to the definite nature of the orders issued to exhibition-patients by the Sequah lecturers.

The tooth-drawing, which helps greatly in also drawing a crowd, while the rapidity of it visibly impresses the spectators, is rather mixed in its results. "Who's dat calling so sweet?" may serve effectually to drown the cries of a sufferer who gets a broken tooth for his pains, but it does not do away with the significant fact that, as we have heard dentists assert, there is always a greater demand for stump-removal after the Sequah lecturer has left a town than there was prior to his visit. Sound teeth must not unfrequently be forfeited, too, in the hurry of extraction. A dental surgeon, practising at the West End, told us that, being at Wimbledon when a Sequah lecturer was performing, he had the curiosity to get nearer, and as a man who had just had a tooth extracted passed from the platform to the crowd he asked the man to let him look at the tooth, which the operator had put into the patient's hand. "Why, it is a sound one," he exclaimed, whereupon such a hostile demonstration was made towards him by some excited and apparently more than ordinarily interested Sequah supporters, that he found it prudent to withdraw himself from the crowd.

CHAPTER VI.

HOLLOWAY'S PILLS AND OINTMENT; SEQUAH'S
"PRAIRIE FLOWER"; LETTER CONCERNING
CLARKE'S BLOOD MIXTURE.

SOME fifty or more years ago a man named Albinolo, one of the greatest of modern French quacks (proud pre-eminence !) issued large numbers of curious little green-covered pamphlets in which he most energetically attacked one Thomas Holloway, of London, and accused him of fradulently appropriating his (Albinolo's) invention. According to this amusing specimen of quarrels among quacks, Albinolo, fired with a notion of creating a fresh field for plunder in England, entrusted Thomas Holloway with the requisite means for bringing out Albinolo's preparations in this country.

This Thomas Holloway seems to have done promptly and thoroughly, but he considerably exceeded instructions, for he conceived the unexpected idea of working them in his own name, and thus aroused the little Frenchman's ire to such an extent that the green cover of Albinolo's pamphlet must have been almost matched by the altered hue of his complexion. " Pity such troubles e'er should come, 'Twixt Tweedledee and Tweedledum "; but it is no part of our province to sympathise with Albinolo, or to side with the (according to Albinolo's asseverations) unfaithful Thomas.

Yet, although Thomas was unfaithful, he did not in one respect wholly partake of the unbelieving character of his scriptural namesake. Thomas Holloway had not achieved absolute success as a tradesman, and he doubtless felt that the time had come for a change of vocation; so that, Albinolo's attacks notwithstanding, he straightway laid in a large stock of drugs, boxes, and gallipots, and ordered a large number of newspaper advertisements. He believed to an enormous extent in the gullibility of the British public, whatever private views he might have entertained as to the universally curative character of his wares; and, moreover, he dubbed himself "Professor" Holloway, jumping at once from his previous modest position behind the counter of his shop in the Strand, near old Temple Bar, to high scientific rank.

His example was contagious, and professors sprang up in all branches of business; so that, in our student time at "Bart.'s" it was quite possible in the course of a few hours to have one's hair cut by a professor, one's measurement for clothes taken by another professor, one's food prepared under the superintendence of a third professor, and one's knowledge of the "noble art of self-defence" improved by so many punches on the nose, administered for a more or less reasonable fee, by a hulking, beetle-browed professor of pugilism. In fact, professors became almost as common in England then as men bearing military titles are in the United States at the present day.

We have no means of ascertaining whether Thomas Holloway added to his assumed accomplishment of Curer-General an acquaintance with the writings of the English poets; though, consciously or unconsciously, he moulded his course in accordance with the following lines from Samuel Butler's *Hudibras* :—

> "To quack of universal cures;
> And mighty heaps of coin increase."

We commend this quotation to the present "professorial" staff at Holloway's New Oxford Street establishment, as more apt and more truthful in import than the quotations which the compiler of their *Family Friend*, presented to customers, has reproduced from Shakespeare and other authors, with more ingeniousness than ingenuousness, for the purpose of puffing Holloway's Pills and Ointment.

As the *Family Friend* gives a number of anecdotes amidst other miscellaneous matter, we naturally searched its pages for some bearing upon the life of the late "professor," but we were doomed to disappointment.

Under the heading of "Seasonable Advice," the pills and ointment are pertinaciously recommended for all ages, all climates, all diseases, all seasons, and in all quantities. We recollect, when on a visit to Wiltshire, noticing a roadside inn at Marlborough bearing the quaint name of the " Five Alls." The device on the sign-board swinging in front of the house is divided into five compartments, representing— 1, the Queen, with the motto, " I govern all " ; 2, 3, and 4, a robed bishop, a general, and a judge, respectively praying for all, fighting for all, and administering the law for all ; while the fifth figure is that of a man, typical of the British tax-payer, who says significantly, " I pay for all." A somewhat similar condition of things exist as regards quack medicines, for the credulous consumers of these nostrums pay for all, whether it be the huge establishments in which the business is carried on, the many thousands of pounds spent in advertising, or the " mighty heaps of coin " which the proprietors amass.

Well, although we were disappointed at not finding any anecdotes of the late " professor " in the publication to which we have referred, we need not on that account leave our readers wholly in a like predicament, but will give two excellent stories which have come to our knowledge.

When Charles Dickens was in the height of his splendid career as a novelist, Holloway sent him a cheque for £1,000, with an intimation that he might consider it as his property if he would insert in an early number of one of his works, then coming out in a serial form, some reference to the Holloway patent medicines. Dickens, to his honour be it said, with equal promptitude and indignation, returned the proffered bribe. Upon hearing of this incident, Thackeray remarked, with the quiet sarcasm of which he was master, that if he had been in Dickens' place he should have killed the villain of the novel with an overdose of Holloway's Pills, and thus have secured the £1,000.

On another occasion, in the year of the Great Exhibition, 1851, there was a large meeting of representative men at Gore House, Kensington. Holloway had gained admission with the throng, and made use of Mr. George Augustus Sala, whom he happened to know, to obtain introductions to prominent personages. He was particularly desirous of being brought under the notice of Thackeray, and Mr. Sala, probably for reasons not difficult to seek, was equally anxious to avoid this. However, yielding at last, Mr. Sala took the "Professor" up to the great novelist, and managed to say a few words of introduction, despite the crowd around them. Thackeray appeared to understand the name imperfectly, and complimented the "Professor" in the same strain as he would have done in the case of a distinguished military officer. Holloway, confused, had to explain that he was not a General, but merely "Professor" Holloway. "Oh! well," observed Thackeray, "I made a very natural mistake, for you, too, must have killed thousands of people."

As to the composition of Holloway's pills, Mons. Dorvault, an eminent French chemist, reported it to be as follows in 144 pills. For convenience we give the amounts in English instead of French weights:—

Aloes	62 grains.
Rhubarb	27 ,,
Saffron	3 ,,
Sulphate of Soda	3 ,,
Pepper	7 ,,

Aloes, which the Sequah quacks falsely describe under the name of "Prairie Flower," is, like rhubarb, a substance possessing aperient properties; saffron and sulphate of soda are in such infinitesimally small quantities that they may be passed over without comment, and the pepper may be similarly dismissed as unworthy of consideration.

Holloway's ointment contains, according to Mons. Dorvault, in 159 parts:—

Olive Oil	$62\frac{1}{2}$ parts.
Lard	50 ,,
Resin	25 ,,
White Wax	$12\frac{1}{2}$,,
Yellow Wax	3 ,,
Turpentine	3 ,,
Spermaceti	3 ,,

There is no ingredient here, any more than in the pills to which special remedial properties can be attributed; yet Holloway's advertisements claim for both preparations marvellous curative characters, and assert in the most unblushing and untruthful manner that cholera, typhoid fever, diphtheria, asthma, pleurisy, influenza, dysentery, gout, rheumatism, all skin affections—in short, every ill to which human flesh is subject—will vanish away upon the administration of the pills and the inunction of the ointment, like morning mists before the rising sun.

"Allah is great!" *Holloway* is greater; that is, if one is foolish enough to believe what Holloway says. Instead of the few grains of pepper in the pills, we should, bearing in mind the old proverb, require many grains of salt before accepting the Professor's professions of perfection.

While on the question of analysis, we may mention that a correspondent informs us that lately a Sequah lecturer in a country town had the brazen impudence to tell his gaping audience that Sequah, Limited, defied the most skilful analyst to find out the composition of the "Prairie Flower" Mixture, and offered to give £1,000 to anyone who could state what it contained. Mr. A. W. Stokes, F.C.S., Public Analyst for Paddington and other important metropolitan districts, has made for us a very complete analysis; but we have not learned that Sequah, Limited, have forwarded a cheque for £1,000 to that gentleman. Possibly they are considering the desirability of repeating the offer; in which case it is to be hoped they will also deposit the money with some person in whom more reliance can be placed than in the wild and wholly untrue assertions of the Company's lecturers, whether as regards the analysis of the Sequah nostrums, or their boasted remedial properties.

WITH reference to our articles on Clarke's Blood Mixture (Chapters II. and III.), we have received the following letter from an old and esteemed contributor, Dr. Bartlett, whose name is well known through his professional position as an analyst and the good work which he did in connection with the passing of the Adulteration of Food and Drugs Act, and also through his subsequent exposure of adulterations.

In common with Dr. Bartlett, and doubtless many of our readers, we look with a large degree of expectancy and interest for the reply to the question put by him to the proprietors of Clarke's Blood Mixture.

To the Editor.

Dear Sir,—After some years' absence from the busy centre of current sanitary and scientific literature in London, I am gratified to find a fresh impulse infused into your useful periodical by its exposures of quackery.

I am particularly interested in the raid made upon patent medicines and other proprietary articles, and it rouses the blood of an old Commissioner of Public Health to find these flagrant abuses still pursuing the profitable tenor of their ways, undisturbed by exposure and almost unchallenged, except in your columns. I am not sure that I am in accord with all the views you have expressed on this subject. I am not even certain that I would, if I could, wipe out of existence every patent medicine, and certainly I should wish to use the greatest discrimination in "sifting out the wheat from the chaff" in dealing with what are termed "proprietary articles."

I would draw the line at those patent medicines which are, from the nature of their composition, likely to be injurious to health; those which must be detrimental from the dosage proposed and recommended by the proprietors; those which pretend to cure certain diseases, and which are demonstrably unsuitable or inert for the purposes for which they are sold; and, lastly, all those concerning which any false pretence is advertised or published, or whose proprietors are unable to authenticate every statement advertised or published by them. Touching on your recent articles, I have had almost thrust upon me a most extraordinary advertisement in to-day's *Daily Chronicle*. It apparently emanates, in the first place, from the *Family Doctor*. There are no marks of quotation to the first fifteen repetitions of the *Family Doctor's* advice (in capitals); then comes a quotation, as I take it, from some very peculiar publication, bearing, ironically, I suppose, the title of the *Family Doctor*. I say ironically, because no respectable family doctor would dream of giving such advice as :—"Clarke's Blood Mixture is a curative agent which cannot be too highly estimated, since it cleanses and clears the blood from all impurities, and restores it to its normal condition. It is a medicine of the greatest possible value. It is certainly the finest blood purifier that science and skill have brought to light up to the present time, and we (the *Family Doctor*) can with the utmost confidence recommend it to our subscribers and the public generally."

This is strong and unctuous enough, and no doubt catches many unsophisticated sufferers from pimples, eruptions, sores, scrofula, scurvy and bad legs, skin and blood diseases of all descriptions, as if with "bird-lime." But having swallowed the

Family Doctor with extreme nausea I come to a "staggerer," as Mr. Dick Swiveller observed. It appears to be a quotation from a certificate signed by my old friend Dr. Alfred Swaine Taylor, stating that "Clarke's Blood Mixture is entirely free from any poison or metallic impregnation, does not contain any injurious ingredient, and is a good, safe, and useful medicine."

I now simply ask the proprietors of Clarke's Blood Mixture to authenticate that quotation with the rest of the text, if any, under Dr. Swaine Taylor's actual signature, giving also the date.

I have very particular reasons for pressing this inquiry, and if the authenticity of the quotation is not sufficiently proved, I will lay these reasons, with other information, before your readers in a future communication. It will be interesting to show how such quotations are obtained and used.

       I am, faithfully yours,

           H. C. BARTLETT.

Chapter VII.

Saved from the Waste Paper Basket; Correspondence concerning Holloway and Mattei.

I.—Concerning Holloway.

To the Editor.   Dublin.

Sir,—I am *out of patience* with such abominable articles as you have inserted in your paper about what you are pleased to call *quack* medicines. Do you not know that the late esteemed *Professor* Holloway was a gentleman of the *highest* character, and that he founded several institutions of *almost national* importance? You may not *know all this* but I do, and I am *proud* to say that I have a near and *dear* relative who has *enjoyed* the benefits of the Holloway College at Egham. And don't you know that one who bears the *honoured* name of Holloway has been knighted by Her Majesty?

Yours *indignantly*,
A Lady.

***

II.—Concerning Mattei.

Another correspondent has sent us two fearfully lengthy epistles from Pisa. Judging from his violent and illogical attacks on the medical profession, his judgment must be as much out of the straight line as the famous Leaning Tower which forms one of the sights of that Italian city. The oddest thing, perhaps, about his letters is that he has adopted the signature "COMMON SENSE." He might as well, while about it, have unconsciously imitated the signature of the "Professor's" lady champion and signed himself "A GENTLEMAN."

These communications are too long for publication in full, and we must therefore content ourselves with extracts. One lump of the mud thrown is so much like another that a sample or two will serve as well as exhibiting the whole heap :—

"Without wishing to appear abusive, I must say such an article as that on Mattei's Remedies, is a reflection on your honesty, common sense, and capacity as editor of a paper.

" 1. *Honesty.* Is it honesty to denounce as sheer humbug a system of which you *know nothing?* You pin your faith to Mr. Stokes's analysis of the ' Electricities,' but you *don't even try the remediés,* and you condemn them beforehand. What good are your assertions without proofs? Prove that they do no good or do harm, and people will listen to you; otherwise they will think you belong to the great conspiracy to silence any new remedy not brought out by the doctors. Let me add, too, that I know a doctor who analysed the ' Electricities,' and *fancied* they were prepared from fungi, to which idea the smell and the growths that devolop in them might give colour.

"2. *Your common sense.* Do you believe your little *ipse dixit* will stop people trying Mattei ? What an idea you must have of the mental capacities of your readers if you imagine you can keep them off the forbidden Mattei territory by calling him a humbug.

"3. *Your commercial capacity as the editor of a paper.** Your journal looks a useful kind of publication, but the line you have adopted would quite discourage myself and others from taking it in." * * * *

"Do you not feel that your article, imputing fraud to Mattei, is abusive? Not only abusive, I should say, but libellous; and abuse naturally generates abuse. I hope my abuse has been restrained by gentlemanly feeling, but it is difficult for *those who know* to write in cold blood to people like yourself, who *knowing simply nothing*, and not even taking the trouble to test a single remedy, sit down in an editor's chair and pronounce an *ex cathedrâ* judgment, that it is all a humbug and a fraud! I wonder you cannot see that a paper conducted on such principles can never go down with people who have a grain of common sense or reflection. We should be thankful to you if you would unmask Mattei by *proving* his fraud; but that is the last thing that occurs to you."

* * * *

"P.S.—I dare say you have never compared your position with that taken up by the Pope. Were the Saviour to come again, and work miracles of healing in your sight, you would pronounce them humbug, if not fraud."

In accordance with the French rule, "*Place aux dames*," we will deal first with "A Lady's" communication. While admitting that her relative may have profited from the posthumous benefits provided by the late Thomas Holloway, we are not prepared to admit that the knighthood conferred on the present head of

---

* "'Commercial capacity'"! We did not know before that commercial capacity was an editorial essential. At the same time we must allow that Mattei could give most people points in commercial capacity, if by that term is implied the impudence requisite for selling water at the rate of about 5s. an ounce.

the firm of pill makers has added to the lustre of that ancient order, that it has elevated quackery into a science, or that it has in any degree altered the fact that Thomas Holloway "made his pile" by the pertinacious and persistent puffery and exaggeration of the remedial value of his wares.

The "almost national" importance of the institutions founded with part of the money thus amassed is a question which we will not stop to discuss, beyond remarking that there was one institution which Thomas Holloway established, not the least appropriate, perhaps, if intended as a delicate compliment to the people who believe that his pills and ointment will cure every human malady, viz., an asylum for lunatics.

After all, our feminine correspondent has advanced no sounder argument in the "Professor's" favour than that used by Lucetta in Shakespeare's *Two Gentlemen of Verona* :—" I have no other but a woman's reason ; I think him so, because I think him so."

Without "wishing to appear abusive," the Matteist advocate has succeeded beyond his desires or expectations—what would he say, we wonder, if he tried to appear abusive?—and we complacently accept his attack as a pretty good proof that the shafts aimed at Mattei's imposition have hit, at least, one of his followers. It would seem that our correspondent looks upon common sense as a rare quality, limited to himself and other Matteists; while he advances the illogical argument that the circumstance of a man having devoted special study to any particular branch of knowledge wholly disqualifies him from coming to any right opinion thereon. How cleverly, too, he works in his similes! Scarcely have we got over the contrast of our humble selves with the Pope, though we never claimed infallibility (we leave such preposterous claims to people of "Common Sense's" stamp), than the artfully simple, or simply artful, comparison of Mattei with the Saviour (!) is adroitly introduced.

As for our knowing nothing of Mattei's remedies, this assumption by "Common Sense" is as incorrect as his other insinuations, for it happens that some years ago our attention was directed to these preparations by General Booth's asking our opinion concerning them, in the course of a conversation about Mrs. Booth's illness with cancer. It was evident to us—as Mr. Stead has confirmed in his panegyric on Mattei in the *Review of Reviews*—that the General had no great faith in the "Electricities"; but we availed ourselves of subsequent opportunities to put the Mattei remedies to practical tests. The results were so unsatisfactory that we were not surprised when Mr. Stokes, after submitting the "Electricities" to rigid analysis, announced that the samples sent to him, just as they were purchased from the London agents, consisted only of water. Water, "honest water," as Shakespeare wrote, to what base uses may you not be brought! And, because we say that so long as Mr. Stokes' analysis is not disproved, and that so long as

Mattei and his followers content themselves with vague assertions we shall regard him as a gross humbug, "Common Sense" avers that we are not only abusive, but libellous. He had better reserve these charges till Mattei has proved that the analysis we published is false, and has shown what his "Electricities" do contain besides water.

There is a story told of a gentleman who, passing along a street of a manufacturing town, was so astonished at the spectacle of a little woman vigorously punching a big "navvy" in the back, at the same time using most angry language, without any attempt at resistance or remonstrance on the part of the recipient of the blows, that he could not refrain from speaking to the man. The good-humoured giant turned round towards the gentleman, and, removing his pipe from his mouth, observed, with a broad smile on his face, glancing down at his diminutive and irascible spouse, "It's all right, sir; it pleases her, and it don't hurt me." In like manner it may please "Common Sense" and others of the same stamp to pitch into and vilify the medical profession, while it certainly need not hurt nor disturb the equanimity of the latter.

We have already referred in these articles to the promptitude with which quacks assert that their nostrums will cure every ill under the sun. Universality is their great aim, as embodied in the advice given by an American patent medicine manufacturer to a relative who was about to embark in a similar business:—"Recommend the stuff internally, externally, and eternally."

Holloway's Pills and Ointment afford a good illustration of the extent to which this advice can be carried. The "boon and blessing to every family," which one sees advertised at all railway stations, consists of aloes and rhubarb, with some still less important constituents, to be administered internally; and an ointment composed of lard, oil, wax, turpentine, and resin, to be used externally; whilst the constant and persistent advertising of these nostrums may be regarded as carrying out the third condition laid down by the successful Yankee quack. Anyone acquainted with even the most elementary rudiments of pharmacy and medicine must, of course, see that there is nothing of a specific—far less of a marvellous—character among the ingredients we have enumerated; yet Holloway's advertisements assert that these two quack preparations will promptly and certainly cure every disease, from asthma right down through the alphabet.

We wonder whether the Bishops and other bigwigs who manage the affairs of Holloway College, referred to in "A Lady's" letter, ever give five minutes' thought to the questionable means and despicable disregard for truth by which the money was amassed that enabled the founder of that institution to leave behind a perpetual memorial of Holloway's Pills and

Ointment and of human gullibility. *Monumentum ære perennius* might in this instance be translated, "A monument more lasting than the brazen assertions of its founder."

Some time ago, when General Booth was raising the special Salvation Army fund, one of his opponents tried to trip him up by asking whether he could conscientiously accept money in aid from unbelievers and other unworthy persons, referring to some whose names had appeared in the list of subscribers. To this the General replied that if the Devil himself should send him a donation he should not hesitate to accept it and apply it to the purpose in view. The somewhat equivocal moral of this is: Blind your eyes as to the source whence the money is derived; and it is apparently fully acted up to by the trustees of the Holloway College, and by the people who, ignoring the mean misrepresentations and dishonourable way in which the funds were accumulated, regard the Holloway College as a memorial of a public benefactor, instead of a permanent proof of human credulity.

Chapter VIII.

Allen's World's Hair Restorer; Mexican Hair Renewer; Singleton's Golden Ointment for the Eyes; Rowland's Kayldor and Gowland's Lotion for the Skin; Anna Ruppert's Skin Tonic; More about Mattei.

In previous articles we have dealt with quack preparations for internal administration. In the present, we will speak of others for external employment. Even the most highly-vaunted and most widely-advertised of these turn out to be of a very ordinary, or—worse still—dangerous composition, when they are submitted to the close scrutiny of the analyst. We give the analyses of some of these:—

1. *Allen's World's Hair Restorer.*—An analysis of the contents of an eight-ounce bottle indicated $75\frac{1}{2}$ grains of sulphur and 87 grains of acetate of lead. Considering the low commercial value of these two drugs, the most expensive, however, to be found in the preparation, the price at which it is sold cannot fail to give a very handsome return to its proprietors. If it were of corresponding value to persons who fondly and foolishly imagine that luxuriant crops of hair

will adorn their smooth, shiny scalps as the results of its persistent application, the charge made for it would, of course, be a different thing altogether.

2. *Mexican Hair Renewer.*—The foregoing remarks apply also to this nostrum, which is as much overrated as over-advertised. It consists of acetate of lead 1 part, precipitated sulphur 4 parts, glycerine 30 parts, and water 160 parts. The only ingredient capable of promoting capillary growth is the glycerine, and that is so completely swamped with water in the proportion of more than five parts to one as to be rendered inert.

3. *Singleton's Golden Ointment for the Eyes.*—Golden only in name, for it contains not the faintest trace of the precious metal. As a matter of fact, it is practically identical with the nitric oxide of mercury ointment of the old pharmacopœia, sometimes efficacious in cases of ophthalmia, but in the majority of instances of eye diseases worse than useless. It is, unfortunately, much easier to tamper with the eyes than it is to remedy the mischief resulting from indiscriminate and injudicious treatment.

4 and 5. *Rowland's Kalydor and Gowland's Lotion for the Skin.*—Mr. Henry Beasley, a good authority, and a voluminous writer on all matters connected with pharmacy, states in his " Druggists' General Receipt Book " that either of these well-known cosmetics, so *dear* to the fashionable and vain amongst our lady friends, may be imitated as follows:—Bitter almonds

(blanched) 1 ounce, corrosive sublimate 8 grains, rose water 16 ounces. In the last (ninth) edition he says that, on more recently examining a sample of Kalydor, he could not find any mercurial ingredient, so that the corrosive sublimate has been wisely—let us hope permanently—eliminated.

6. *Anna Ruppert's Skin Tonic.*—But if the makers of Kalydor have arrived—tardily, it must be admitted —at a consciousness of the dangerous character of so powerful a drug as corrosive sublimate (bichloride of mercury), it seems that there is at any rate one cosmetic vendor who is absolutely reckless in this respect—a Mrs. Anna Ruppert, self-styled the "celebrated American Complexion Specialist." Now this individual does not hesitate to put as much as a grain of corrosive sublimate to every ounce of a largely-advertised lotion for the skin. So Dr. B. H. Paul, the well-known analyst, says in his report upon the composition of a specimen of this lotion submitted to him for examination; and he adds the significant observation that the specimen submitted to him was "weaker than other samples previously examined," so that the manufacture must be conducted in such a careless, haphazard way that the actual quantity of this powerful poison would appear to be left to chance.

A leading medical paper, the *British Medical Journal*, gave, some little time back, the details of a case of acute inflammation of the jaw from mercurial poisoning, reported by Dr. W. H. C. Staveley, F.R.C.S.,

L.R.C.P., Lond., and Mr. R. Denison Pedley, L.D.S., F.R.C.S., Edin. The patient, a widow lady, rather over thirty years of age, was first seen by Dr. Staveley, who recommended her to consult her dentist on account of her teeth. She was suffering from great weakness, boils, and severe pains in the upper and lower jaw; the teeth were all loosened (several having to be removed in consequence of this condition and of abscesses forming at their base), and the bony structure of the jaws was much inflamed and thickened. After the lady had been under treatment for some time, her sister informed Dr. Staveley that the patient had for three months previous been using a lotion for her complexion, and that the symptoms had commenced shortly after she began to use it, becoming gradually intensified until she was compelled to seek medical advice. The lotion turned out to be Mrs. Anna Ruppert's special Skin Tonic, and the large proportion of corrosive sublimate found in it by Dr. Paul, F.C.S., clearly demonstrated the cause of the mercurial poisoning, the severe, continuous pain, the injury to the teeth, and all the other alarming symptoms.

We make the following extract from the lady's own statement:—" In October, 1892, I went to Mrs. Anna Ruppert, in Regent Street, my complexion not being good; otherwise I was in good health. She looked at my face with a magnifying glass, and said that her preparation would remove spots. She recommended me to try her special treatment. Next morning, she sent me by parcel post three bottles of lotion, three cakes of soap, one pot of ointment, and one box of powder, for which I paid two guineas. The interview lasted about five minutes. About a month after I had been to her I began to be sleepless, lost my appetite, and my hands became so tremulous that several of my friends noticed it." The remainder of her statement deals with the various symptoms which have already been detailed. Had she not, fortunately, consulted a medical man when she did, there is little doubt that, by continuing the use of this lotion, the lady might have died of mercurial poisoning. As it is she has suffered months of agony and misery, and her health is impaired.

The poor victim has certainly had a great deal to show for the two guineas which she foolishly handed over to Anna Ruppert for "special treatment." A pamphlet which one of our lady readers has sent us, entitled "The Book of Beauty," by Anna Ruppert, of the usual catchpenny style of such publications, states that "the skin tonic is not in the least to be feared," from which one would be led to infer that the laws of nature are suspended when the "special treatment" is adopted, and that in Anna Ruppert's hands the most powerful mineral poisons are miraculously rendered absolutely innocuous. Still, we would almost as readily believe this as we would the assertion that it "is endorsed by the best living physicians." If

Anna Ruppert had not occupied so much of her pamphlet with the most glowing praises of herself, she might have found room for the names of these "best living"—wonderfully ignorant and credulous individuals, whoever they may be, presuming that they exist outside her pamphlet. Among other information which she does give is a price-list of her "specialities":—Skin tonic, single bottles, 10s. 6d.; three together, 25s., usually required. What for? A less quantity would produce mercurial poisoning, according to the unfortunate widow's experience. Dyspepsia Cure and Liver Regulator, 4s. 6d., and so on.

We have stated that Mrs. Anna Ruppert can be "consulted" at Regent Street, London, but in an advertisement taken from a ladies' journal, no less than fourteen other addresses are given at Brighton, Edinburgh, Dublin, Manchester, and other large British towns, as well as at Paris, Berlin, Vienna, Calcutta, Sydney, &c. Sir Boyle Roche's bird, which, according to the eccentric baronet, was so swift in its flight that it could be in two places at the same time, sinks into insignificance as compared with a complexion specialist who can be "consulted" at no fewer than fifteen different places.

Perhaps some of these addresses have slipped in in error. Our reason for this suggestion is that Berlin is included. It is scarcely probable that Mrs. Anna Ruppert would regard this city as a happy hunting ground, seeing that some time back she visited Berlin, and advertised her "specialities." The municipal authorities there have a quick method of dealing with such cases, and the President of Police for Berlin cut short Mrs. Anna Ruppert's stay in the German capital by the issue of a public caution, of which the following is a literal translation:—

PUBLIC NOTICE.

As a cure for the most varied skin diseases, a Mrs. Anna Ruppert recommends her Skin Tonic in the daily newspapers. This secret remedy consists of a solution of corrosive sublimate in water, with the addition of some glycerine, and it is slightly perfumed. It is sold in bottles containing about five ounces for the sum of eleven shillings, while the real value of the bottle's contents is less than three farthings. This announcement is hereby given in order to warn the public.

THE PRESIDENT OF POLICE,
Berlin.

Three farthings worth of stuff for eleven shillings! There is little room for wonder that Mrs. Anna Ruppert, who assures her readers that she knows her "profession" thoroughly, promises to "cheerfully give" advice to all comers, when she can get such a grand return for so trivial an outlay.

OH, DEAR! WHAT CAN THE *MATTEI* BE?

We have received the following registered letter.

COUNT MATTEI'S REMEDIES.
Central Depôt for the United Kingdom and Colonies,
London, July 5th, 1893.
To the Editor.

Sir,—My attention has just been called to the fact that in

your issue of May 13th you reprinted one of the articles in which you attack Count Mattei.

I have now to inform you that the Electricities have been analysed by two other analysts, and that if, in the face of their analyses, you continue publishing your attacks upon what have been proved by competent medical authorities to be genuine remedial agents of a very valuable character, I shall have to hold you responsible. I enclose herewith a copy of our monthly publication for May 15th, in which you will find an article dealing with the question of the analyses, to which I beg you to give careful consideration; also one for November 15th, 1892.

Yours faithfully,

A. GLIDDON, Manager.

A refreshingly cool letter, certainly, with thermometers bursting around us in their futile efforts to record the high temperature that will make the summer of 1893 remembered many years after the Matteist craze has followed the example of the thermometers; but if the Matteist manager imagines that we are to be muzzled so easily, he is labouring under as great a delusion as many of his customers—some of whom talk such utter nonsense as that the Electricities will cure broken limbs, and restore sight to the blind, not to mention other alleged miracles.

We have to express our thanks to Mr. Gliddon for one act of thoughtfulness on his part, namely, sending only one of the two promised numbers of *Modern Medicine*, which would be more appropriately entitled *Modern Magic*. If ever anything should be taken in homœpathic quantities, it should be Matteist literature, which might be described in the Matteist jargon as *Emetico*.

It appears from *Modern Medicine* for May 15th that the analysts referred to in Mr. Gliddon's letter are M. Casali, of Bologna, and an English chemist, Mr. Butterfield. M. Casali reports that he has carefully examined the Electricities, and submitted them to a great variety of chemical tests. "As a result of these, he had come to the conclusion that the liquids had a characteristic odour which was rendered more marked by heat. This odour or aroma seemed to him to bear a slight resemblance to that of olibanum, especially to that of drops of translucid Indian incense. He felt himself justified in asserting (1) that the Electricities are not simply distilled and pure water; (2) that they have the characteristics of pharmaceutical solutions, *i.e.*, of waters charged or saturated by distillation with the volatile medicinal principles of flowers, leaves, and plants."

As for Mr. Butterfield, *Modern Medicine* says:— "His conclusions agree with those of Professor Casali, so far at least as the presence of an aromatic essence in the Electricities is concerned. Like the Italian professor, the English analyst has been unable to ascertain the exact nature of this essence. He is, however, quite confident that it is a mistake to assert that the Electricities are ordinary water."

Two more bald, jejune, and unsatisfactory analytical reports we never saw, and two hundred or two thousand such inanities could not be regarded by any sensible person as affording any disproof of Mr. Stokes' analysis. Mr. Stokes' report was confirmed to the letter by the independent examination made last June by professor Michaud, chief of the Government Laboratory at Geneva, and both of these analysts assert that Mattei's Electricities had no colour, flavour, or odour, and that the chemical composition is identical with that of water. Perhaps if it had occurred to either of them to have, say a peppermint lozenge in his mouth when smelling the fluids, there might have been an aroma as of "translucid Indian essence." Mr. Butterfield is confident on one point only, viz., that the Electricities are not ordinary water. In a sense, very different, however, to his, we agree with him. Ordinary water can be obtained of any water company at less than 1s. per 1,000 gallons; the Electricities are sold at the rate of 5s. an ounce.

Grievously disappointed with the Matteist analysts, we turned over the pages of the Matteist monthly periodical in the vain hope of discovering the names of some of the competent medical authorities whom Mr. Gliddon holds *in terrorem* over us. We found cases of remarkable cures effected by Madame Schmid, Dr. E. P. (Topeka, Kansas, U.S.A.), Dr. C. (another American, too modest to give his full name), Pastor Stocker (of Mingolsheim), M. (Dunedin, New Zealand), who talks of seeing patients in his "shop":

Margaret Wilson, M. H. W. (senior curate of C.), M. de Rozehuba, X., and Baboo Radhica De (of Calcutta), who is reported to have cured an individual rejoicing in the name of Nibin Chundergaugoolychinsuran. The use of initials only would have been excusable here. It is almost needless to remark that Nibin (we will not goad our unoffending compositors to desperation by repeating the cognomen) is represented to have been "cured like miracle" of inflammation of the eyes by the use of Mattei's Electricities, in less time than any one of our many thousands of readers could pronounce the name.

If these ten electro-homœpathic practitioners, who occupy considerably more than half the space of *Modern Medicine*—two women, two Americans (vaguely designated "Dr." and, with singular modesty, contrasted with their assertions, indicated by initials only), two nondescripts (M. de Rozehuba and X.— certainly not Y. Z.*), one foreign pastor, one English curate, one New Zealand shopkeeper, and one baboo— are to be regarded as fair samples of Mr. Gliddon's "competent medical authorities," they are even more unsatisfactory and unconvincing than his brace of analysts; and that is saying a great deal.

* Wise Head.

CHAPTER IX.

QUACK ADVERTISEMENTS AND TESTIMONIALS;
MOTHER SEIGEL'S SYRUP.

AN American quack, who was ostentatiously boasting of his wealth in one of the chief hotels at Saratoga Springs, was addressed by an admiring auditor as follows:—"There must be a fortune in patent medicines." "I don't know," reflectively observed the quack, "It isn't all profit—bottles cost money." Many persons would probably have added, "and advertising" after "bottles." But the quack knew better than to mention this item, for he might as well have included his own mendacious tongue. He could not reckon advertising in the estimate of cost, for the plain reason that the more he spent in puffing his wares the greater would be the return in cash.

The amount of money spent on advertisements of quack medicines is astonishing. Thousands, tens of thousands, and hundreds of thousands of pounds are unhesitatingly launched upon this expenditure, the advertisers feeling certain of a profitable return for their outlay, and trading on the blind credulity of people who pin their faith on quack medicines, merely because they have read some glowing advertisement (more or less false).

The Anglo-Saxon race used to be regarded as the most easily duped in this respect—the readiest swallowers alike of quack advertisements and quack medicines. But the "heathen Chinee" runs the Englishman and American close in this respect; and, according to an article in the *Cornhill Magazine*, Chinese papers contain even a larger proportion of quack advertisements than English or American periodicals. Moreover, the ingenuity of the "Chinee," when he turns "quackee," puts "the barbarian devils," as he would doubtless term his European or American charlatan *confrères*, completely in the shade. They can only puff, while the Chinamen blow whole gales of lying assertions.

Take for instance, an announcement in the *Shên Pao*, or Shanghai *Gazette*, of the grand "Fairy Recipe for Lengthening Life." "This recipe," says the advertisement, "has come down to us from a physician of the Ding Dynasty. A certain official (this *certain* official is somewhat uncertain) was journeying in the hill country when he saw a woman passing southwards over the mountains as if flying. (The name of this wonderful 'record breaker' is, unfortunately, not given. Could she have been a Chinese 'Mother Seigel'? In her hand she held a stick, and she was pursuing an old fellow of a hundred years. The mandarin asked the woman,

'Why do you beat that old man?' 'He is my grandson,' she answered. 'I am five hundred years old, and he is only one hundred and eleven; he will not properly take his medicine, and therefore I am beating him.' The mandarin alighted from his horse, and knelt down and did obeisance to her, saying, 'Give me, I pray you, this drug, that I may hand it down to posterity for the salvation of mankind.' Hence it got its name," adds the veracious Chinese quack.

Whatever may be thought of the existence amongst moderns of lineal descendants of Ananias, no one can, after this, express any doubt as to the probability that a contingent of them must have found their way across the Asiatic continent to the land of the Celestials.

Reading further, we learn that this fairy medicine "will cure all affections of the five intestines and derangements of the seven emotions" — Chinese physiology is as wonderful as Chinese physic—and that it will speedily and effectually relieve every ailment under the sun. Here are directions for its administration, the dose being equal to a quarter of an ounce :—"Take it for five days, and the body will feel light; take it for ten days, and your spirits will become brisk; for twenty days, and the voice will be strong and clear, and the hands and feet supple; for one year, and white hairs will become black again, and you will move as though flying. Take it constantly and all troubles will vanish, and you will pass a long life without growing old." The price per bottle—our readers must not put the value of this elixir at too many thousands of pounds—is about 3s. 3d. in English money. After this specimen of quack announcements we shall expect to hear of a rush on the part of British patent medicine vendors to secure Chinese managers for their advertising departments, to spread their fame and increase the sale of their pills, potions, and plasters.

There are pessimists who insist upon the fact that there is no time like the past, that men of the present day are degenerate, physically and mentally, when compared with those of bygone times. Certainly, in the matter of hard lying the quacks of other days could almost give points, judging by four advertisements taken from a newspaper published in the early part of the last century.

The first is "an incomparable pleasant tincture to restore the sense of smelling, though lost for many years, a few drops of which (the tincture we presume), being snuffed up the nose, infallibly (at what date, we wonder, was this word first adopted by quacks?) cure those who have lost their smell, let it (the loss is meant, we suppose) proceed from what cause soever." This marvellous stuff, at 2s. 6d. a bottle, was to be obtained only at Mr. Payne's toyshop (at the period of which we are writing toyshops were not as now limited to the sale of toys, but were a kind of bazaar or emporium for many articles) at the

"Angel and Crown, in St. Paul's Churchyard, near Cheapside."

No. 2 was "an admirable confect, which assuredly cures stuttering or stammering in children or grown persons, though ever so bad, causing them to speak distinct and freely without any trouble or difficulty." The advertisements concludes with :—" Its stupendous effects are really wonderful," not omitting to mention that the confect can be secured, at the moderate charge of half-a-crown a pot, at Mr. Osborn's toyshop, at the Rose and Crown, under St. Dunstan's Church, Fleet Street.

The third preparation is a specific for "loss of memory or forgetfulness, certainly cured by a grateful electuary, peculiarly adapted for that end; it strikes at the prime cause, which few apprehend, of forgetfulness, makes the head clear and easy, the spirits free, active, and undisturbed, corroborates and revives all the noble faculties of the soul, such as thought, judgment, reason, and memory, which last in particular it so strengthens as to render that faculty exceedingly quick and good beyond imagination, thereby enabling those whose memory was before almost totally lost to remember the minutest circumstance of their affairs, &c., to a wonder!" Mr. Payne retailed this miracle at 2s. 6d. a pot. What a pity it seems that it cannot now be got for love or money in these times of fierce competitive examinations! What an excellent and delicately suggestive present it would have been at

Christmas, accompanying bills and "accounts delivered" sent out to long-winded debtors!

No. 4 advertisement is that of "an assured cure for leanness, which proceeds from a cause which few know, but easily removed by an unparalleled specific tincture, which fortifies the stomach, purifies the blood, takes off fretfulness of the mind, occasions rest and easy sleep, and as certainly disposes and causes the body to thrive and become plump and fleshy, if (was this *if* a sign of some mistrustfulness, or an artful bit of hedging?) no manifest distemper afflicts the patients, as water will quench fire. It is also the best remedy in nature for all chronic diseases that take their rise from a bad digestion in the stomach, which this specific tincture infallibly rectifies and thereby cures. It is pleasant to taste"—so are many undoubted dietetic cures for leanness—and that universal philanthropist of the last century, Payne, literally gave it away at his toyshop, with directions for use, charging only the ridiculously inadequate sum of 3s. 6d. a bottle!

A reference to the pages of the London Post Office Directory discloses the fact that innumerable societies, occupying columns of small type, exist for almost every charitable object, dispatching missionaries here, there, and everywhere, sending pocket-handkerchiefs to the Polynesians, nicknacks to New Guinea, samplers to Sarawak, lanterns to Labuan, female doctors to the Fijians, and so on;

surely a sufficient sum could be raised among the credulous to constitute an exploration fund, so that a search might be organized for Payne's prescription books. As our contribution, we would give a "guinea" (box of Beecham's Pills, having had it returned on our hands by our analyst, after examining it).

"There is nothing new under the sun," exclaimed the ancient philospher; and "History repeats itself" has grown into a recognised axiom. It is, consequently, no matter of surprise to find that a modern counterpart of what for distinction's sake we will call Payne's No. 4 exists in the widely advertised nostrum known as Mother Seigel's Syrup. Payne asserts that "bad digestion of the stomach" is the root of all diseases; while Mother Seigel, or rather the proprietary trading in that name, insists, in a pamphlet lying before us, that "there is only *one* real disease, indigestion and dyspepsia." We do not wish to be too censorious, otherwise we might suggest that the conjunction *and* in this quotation destroys the force of the remark that "there is only *one* real disease" (*one* in italics, too), but perhaps the author of the pamphlet meant to say, "*or*." Whichever he or she intended to say, we say, "*Humbug.*"

Elsewhere in the same pamphlet the writer speaks of "the intelligent persons who own these medicines." We have not the same opportunity—nor do we desire it—of examining into their or their writer's intelligence, that we have of analysing their stuff; but, in face of the positive assertion that there is only one disease, we cannot wonder or complain, as the writer appears to do, that "there is more or less of an impression upon the minds of some people" (the intelligent owners excepted, of course) "that statements as to the merits and effects of popular medicines should be received with a degree of allowance." Very prettily put, we must observe.

Now, when it is so positively asserted that there is only one disease, is it quite consistent with such a statement to issue with each bottle of syrup a broadsheet containing an account of "the strange and prevailing disease of this country," printed in so many languages, from Arabic down to Turkish, that "all countries" should have been nearer the mark? "Prevailing" certainly conveys the idea that from the writer's point of view other diseases may and do exist, though not to the same extent as that under discussion. Further, after enumerating a whole host of symptoms, "all in turn present,"—enough, as an American would say, to make a man mad on swallowing any nostrum, however nauseous, so that he may escape at least some of them—the writer goes on to say that "medical men have mistaken the nature of this disease. Some have treated it for" (query, '*as*,') "a liver complaint, some for dyspepsia, others for kidney disease, &c., but none of the various kinds of treatment have been attended with success."

None! Well, that is a poser; for if, as the writer

has elsewhere told us, there is only one disease, and that disease dyspepsia, it is singular that medical men should all have failed in their diagnosis when they treated the patient—not the disease, as the writer has put it—for dyspepsia. But a ray of intelligence has just darted into our mind, although we are not proprietors of any patent medicines. Perhaps the writer is not so particular, after all, in his desire to inculcate the doctrine that there is only one disease as he is to impress upon his readers that there is only one remedy for all diseases, and that that remedy is Seigel's Syrup. The medical men whom he has so unsparingly and indiscriminately attacked, in the cause of truth (and Seigel), omitted to give this remedy to their unfortunate (?) patients; hence, he wishes Seigel's customers to infer, arises the failure to effect a cure in any single instance.

Our readers must be getting into a state of feverish anxiety to learn the composition of this wonderful medicine; this "Nature's secret," as the Seigel pamphleteer modestly styles it in the panegyric with which he introduces a number of testimonials from persons who, "of their own accord" (the pamphlet does not enlighten us as to whether also at their own expense), have come forward to make declarations before magistrates or commissioners (not such a remarkable or difficult thing) "with no other motive than the noble one of letting other sufferers know where help is to be had in the time of need."

Mr. Stokes, public analyst for Paddington and other important metropolitan districts, has unveiled "Nature's secret," and we present his report:—

Analytical Laboratory, Vestry Hall,
Paddington Green, W.

Dear Sir,—On September 7th I received from you a sample of "Mother Seigel's Syrup." This was in a four ounce bottle, in its unopened wrappers, and still sealed with the unbroken stamp of the Inland Revenue bearing the words, "A. J. White, Limited, London."

This sample I have now carefully examined, chemically and microscopically.

I find it to be a complex mixture containing treacle, borax, aloes, capsicum, and liquorice.

The active ingredient is aloes, of which I extracted from the 4 oz. mixture the quantity of 120 grains.

I remain, yours faithfully,
ALF. W. STOKES, F.C.S., F.I.C.,
Public Analyst.

Aloes to the right of them, *aloes* to the left of them, *Aloes*——

We must apologise to our readers for thus abruptly bursting into a parody of Tennyson's beautiful verses. Our emotion was for the moment too great to be expressed in plain prose. Perhaps, too, writing on patent medicines brought the Valley of Death of Tennyson's poem into our mind. We had just come across an old acquaintance, whose name has repeatedly occurred in this series of articles; we will not say a *dear* old acquaintance, as aloes is not an expensive drug. Moreover, we felt that we were getting nearer to the discovery of the Philosopher's Stone, the universal panacea as offered to suffering humanity by the principal patent medicine proprietors, "intelligent persons," without doubt (*vide* the Seigel pamphlet). If we must not break out into verse, we would beg at least the favour of being permitted to indulge in capitals. ALOES, "in the name of the *profit*, ALOES," ALOES heads the poll.

As those of our readers who have perused our previous articles will remember, aloes enters largely into Holloway's pills; under the *alias* of "Prairie Flower," it forms the chief component of Sequah's mixture; and now we learn, on the authority of Mr. Stokes, that aloes is the active ingredient of Mother Seigel's Syrup.

We may here note a peculiar fact which scarcely goes to support the "intelligent" theory of the Seigel pamphleteer, namely, that there seems to exist in the minds of patent medicine proprietors a delusion that the growth of aloes is confined to the United States, whereas nearly the whole supply is imported from the West Indies.

It is a curious circumstance that when a common commercial substance like aloes is required for manufacturing a quack nostrum, such great care and expense are alleged to be necessary for the cultivation of the plant and the preparation of the drug. The makers of the Sequah Prairie Flower Mixture profess to ransack the extensive woods and plains of the Far

West in search of the vegetable extract (aloes) used in its composition; and the proprietors of Seigel's Syrup print on their wrappers the following statement:—"The manufacturers of this medicine have been for fifty years the largest gatherers of roots, barks, and herbs in the world. Their botanical gardens are the most extensive in America." Why, one may reasonably inquire, do these "intelligent persons" take the trouble to gather and grow thus largely such simples as aloes, capsicum, treacle, and liquorice, which could be readily purchased in tons of any wholesale druggist? And, by the bye, we would remark that, notwithstanding a fair knowledge of the United States, as the result both of personal travel and of reading, we are absolutely ignorant of the locality in which Mother Seigel's extensive botanical gardens are situated. The address (America) given on the wrappers is decidedly as much too vague as many of the assertions thereon are much too positive; and we have no more belief in their existence than in the other statements put forward by the proprietors of Seigel's Syrup.

CHAPTER X.

CLARKE'S BLOOD MIXTURE; THE ALLEGED TESTIMONIAL FROM THE LATE DR. SWAINE TAYLOR, F.R.S.; THE OBVERSE AND THE REVERSE.

IN Chapter VI. we published a letter from Dr. H. C. Bartlett pointedly drawing attention to the circumstance that in the newspaper advertisements of Clarke's Blood Mixture the proprietors were making extensive use of a testimonial alleged to have emanated from the late Dr. Swaine Taylor, who for many years occupied a prominent position in the analytical and scientific world. We had previously noticed this testimonial in print, and had formed our own opinion concerning its authenticity, an opinion which we found that we shared in common with Dr. Bartlett and others who had enjoyed the confidence and friendship of Dr. Swaine Taylor during this gentleman's lifetime. No one who knew him could possibly be brought to believe for a single moment that he could ever have drawn up such a testimonial—not indeed, in favour of any advertised nostrum, but still less with reference to Clarke's Blood Mixture. Before proceeding further, we will give two quotations

—one known to be in Dr. Swaine Taylor's own handwriting, the other taken from a gigantic advertisement of the Blood Mixture which has appeared very extensively in the daily and weekly newspapers for some time back.

The Obverse.

(From the Report and Analysis of Clarke's Blood Mixture, published by Dr. Taylor in the *Lancet* of 1875, under the heading of "Quack Medicines.")

After describing the ingredients (see Chapter II.), Dr. Taylor wrote :—" Why such a mixture as this should be designated a 'blood mixture' and a 'blood purifier' is incomprehensible. It has no more claim to this title than nitre, common salt, sal ammoniac, or other saline medicines which operate on and through the blood by absorption. Its properties (*i.e.*, those of iodide of potassium, the chief ingredient) are well known, and there is no novelty in its employment. The only novelty in this form of mixture is that the iodide is dissolved in water coloured with burnt sugar, and that it is described as a 'blood purifier.' The four doses directed to be taken daily represents sixteen grains of iodide of potassium, and if the patient taking it is not under medical observation such a daily quantity as this may accumulate in the system and do mischief. In some constitutions the iodide of potassium frequently taken proves specially injurious. It produces iodism."

We may mention here that iodism is the condition

in which symptoms of poisoning of the system show themselves, very similar in character to the salivation, &c., observed in cases of gradual mercurial poisoning. Very often even small doses, such as a grain, if repeated several times a day, will occasion serious effects. Indeed, as a high authority on the action of drugs says, "Iodide of potassium sometimes produces distressing depression of mind and body. The patient becomes irritable, dejected, listless, and wretched. Exercise soon produces fatigue, and perhaps fainting." The same writer also remarks, "A grain, or even less, may affect the stomach." Yet the printed directions enclosed with each bottle, after recommending this preparation as a never-failing cure for a whole host of diseases, actually state that it is "warranted free from anything injurious to the most delicate constitution of either sex."

But we are not now reporting upon the composition and nature of Clarke's Blood Mixture; we have done that at full length already, and what we have under examination at the present moment is the question of the genuineness of the alleged Swaine Taylor testimonial. We will therefore quote what we may term

The Reverse.

(Cutting from an advertisement of Clarke's Blood Mixture, contained in a London daily paper, August 20th, 1893.)

"CLARKE'S BLOOD MIXTURE is entirely free from any poison or metallic impregnation, does not contain any injurious ingredient, and is a good, safe, and useful medicine."—ALFRED SWAINE TAYLOR, M.D., F.R.S., Lecturer on Medical Jurisprudence and Toxicology.

It is absolutely impossible to reconcile these two diametrically opposite statements. The Obverse, known to have been written by Dr. Swaine Taylor, embodies the deliberate expression of a man of science, the words are well weighed, and their writer is evidently prepared to prove up to the hilt, everything that he has said. But when we turn to the Reverse, what do we find? A positive denial of everything that Dr. Swaine Taylor had previously published, and that had gone uncontradicted for the simple reason that it was all true. Continuing our metaphor of a coin, though the Obverse is of good sterling metal, the Reverse is too brassy to be allowed to pass. If the coin were actual instead of metaphorical, the Reverse would speedily insure for it the fate of the spurious bits of metal that one sometimes sees in a village shop nailed to the counter as a fictitious sham and a warning to evil-doers. Is there any human being outside Colney Hatch or Hanwell Lunatic Asylums who would attempt to uphold the argument that the same man who wrote the first opinion also penned the second? Scientific men of high reputation and honourable position do not "turn about and wheel about" in this Jim Crow fashion. Of all with whom we have been acquainted during a long experience, we cannot point to one less likely to "give himself away" in this incomprehensible style than the late Dr. Swaine Taylor.

In a former number of the journal from which these articles are reprinted, we put the following questions to the proprietors of Clarke's Blood Mixture:—1. When, where, and under what circumstances did Dr. Swaine Taylor give the alleged testimonial? 2. By whom was his signature witnessed? 3. When and where can the original be inspected? To these questions we would add yet two others, namely— 4. Why did the proprietors of the blood mixture withhold from public knowledge so important a testimonial until years after Dr. Taylor's death? And 5. How do they account for using it so extensively directly after we first published his undoubted and unchallenged opinion, in our article on Clarke's Blood Mixture?

They have had a long interval in which to answer the first three queries, but they have remained carefully dumb. Having promised Dr. Bartlett in his fatal illness that we would follow this matter up, we shall not shrink from fulfilling our undertaking. On July 24th, only six days before his decease, Dr. Bartlett dictated a letter which was forwarded to us, in which he said that, feeling that the hand of death was upon him, he must try to complete earlier than he had intended the evidence which he was able to give concerning the alleged certificate advertised by the proprietors of Clarke's Blood Mixture. He goes on to relate that he was standing in the office of a gentleman whom he names, and that, Mr. Clarke happening to come in, this gentleman began to tease Mr. Clarke about Dr. Taylor's genuine report. "Mr. Clarke laughed and said, 'I shall wait a few years till the old fogey is dead, and then no one can prove that he did not give me a certificate.' Shortly after this" (continues Dr. Bartlett, in what might almost be termed his dying deposition) "at one of the final meetings of the Arsenical Wall-Paper Committee of the Society of Arts, I met my dear old friend (Dr. Swaine Taylor). I simply told him what Mr. Clarke had said. He was horrified, and exclaimed, 'Defend me if he should carry out his threat, and you may say that I never did, and never should, give a certificate for any such article, and I certainly never gave one for Clarke's Blood Mixture.' If the condition I am now in," concludes Dr. Bartlett, "adds anything in solemnity to the above-given plain statement of facts, let it be so; but I have now carried out my duty to my old friend, Dr. Alfred Swaine Taylor." (Signed) H. C. BARTLETT.

As Shakespeare wrote, "the tongues of dying men enforce attention." We can draw but one inference from all the circumstances of the case; and we will leave our readers to form their own conclusions, which will doubtless coincide with those to which we have been irresistibly impelled.

Chapter XI.

Quack Testimonials; Clarke's Blood Mixture; Mother Seigel's Syrup.

The people who "quack of universal cures," in the hope that they may "mighty heaps of coin increase" by this practice (*Hudibras*, Part III., Canto 1), are not over particular as to the means adopted, the end being the object which they keep steadily in view; but, at any rate, common prudence should dictate to them the desirability of having a semblance of truth in the testimonials they print and circulate. For instance, in a weekly paper lying before us (dated August 20th) we find the proprietors of Clarke's Blood Mixture boasting in a whole column advertisement that they have the "largest sale of any medicine in the world" and that they possess "thousands of testimonials"; a circumstance to which they apparently attach so much value that they relieve their pent-up feelings of pride by repeating the alleged fact three times in capital letters.

If all the testimonials which the proprietors of the blood mixture make such a parade of are like that we have printed in Chapter X., then we say that they must be the possessors of the largest collection of forged documents ever brought together.

Our five questions also printed in Chapter X., have not elicited any reply, and we did not expect that they would, for the simple reason that they are unanswerable. We gave more than sufficient grounds for the conclusions we have arrived at, and these are: —(1) That Dr. Swaine Taylor never gave the alleged testimonial; (2) that, consequently, no one ever witnessed his signature; (3) that, either no such testimonial exists, or if there be any such document in existence, inspection would prove it to be an unmistakable forgery. In short, the name of a scientific man of high reputation and of undoubted integrity has been used for base purposes after his death.

Sometimes, in their haste, patent medicine vendors commit the error of treating the doctor as dead while he is yet alive. For instance, several years back, the proprietors of the quack remedy called Mother Seigel's Syrup largely advertised a testimonial from a railway guard, in which they made very free use of the name of a well-known Manchester medical practitioner, who was wrongly described—not the only false thing in the guard's testimonial, by the way—as the late Dr. Dacre Fox. But the *late* Dr. Dacre Fox turned up most inconveniently, for, instead of being dead, he had only changed his residence. He brought an action for libel againt the Seigel's Syrup proprietors, which was heard before Mr. Justice Lawrance at the Leeds Assizes, with the result that the jury awarded him

£1,000 damages. The defendants obtained stay of execution only by paying £1,000 into court pending an appeal; and they were no more successful in London than at Leeds. The judges sitting in the Court of Appeal were the Lord Justices Lindley, Bowen, and Kay. On the conclusion of the arguments Lord Justice Lindley, in giving judgment, commented severely on the conduct of the defendants. It certainly appeared to him, Lord Justice Lindley observed, that the defendants had published the libel under the impression that the plaintiff was really dead, and that they would be perfectly safe, inasmuch as the person libelled was dead, and could not turn up against them. Unfortunately for them, Dr. Fox was alive, and instituted an action. The defendants tried to justify the libel, which they unquestionably failed to do. There was not a tittle of evidence to show that Dr. Fox had been guilty of the conduct which they imputed to him. The whole object of the libel was to puff the defendants' wares, and they went out of their way to libel the plaintiff for the purpose of puffing their own quack medicines. They were utterly unscrupulous as to the means taken by them, so unscrupulous that they did not shrink from casting blame upon a person supposed to be dead. The Court saw no reason to grant either a new trial or reduce the damages; therefore the appeal would be dismissed with costs. Lord Justices Bowen and Kay concurred in this decision.

Chapter XII.

Beecham's Pills.

An old Scotch proverb runs as follows:—"There is but one good wife in the world, and every man thinks he has her." This is a curious proverb, which, like most of its class, admits of a double rendering; for, either good wives must be a rare commodity, or Scotch husbands must be far more uxorious and credulous than might be expected of the residents of a nothern clime. If a patent medicine man had the making of proverbs he would probably modify the Scotch saying into "There is but one remedy, and every quack asserts that he is the sole possessor of it." It could not, however, be said that he treasured it up, but that his chief aim was to part with it, even (as he would have his customers believe) at a most alarming sacrifice.

Take, as an example, the pills so widely advertised, and as equally loudly puffed by their manufacturer, Thomas Beecham. "Worth a guinea a box," he declares them to be, with such persevering pertinacity that one is almost compelled to imagine that Thomas Beecham is the very antithesis of his scriptural namesake, or that he had made the assertion so often that he has finally come to believe in it himself. Yet he is ready to sell any number of boxes at the ridiculously small sum of one shilling and three-halfpence, which trade discounts would further reduce to ninepence or less. We need not be too particular on this point, but will assume, for the sake of a simple calculation, that Beecham receives for every box that leaves his establishment the grand total of nine bronze pennies.

Just now we quoted a proverb, and the remarks made in the previous paragraph remind us of another, which lays down the commercial axiom that "a nimble ninepence is better than a slow shilling"; implying that it is better to do a quick turnover at small profits than a slow one showing a larger profit on each transaction. Yet whoever heard of such a straining of this proverb as "a nimble ninepence is better than a slow guinea"? But Beecham does not get even ninepence per box, for the sum of three half-pence vanishes in connection with the medicine stamp that adorns each box, and, as will be shown shortly, is the most costly part of the business.

Was such reckless trading ever known before? The only instance which can at all compare with it is that of the old apple-woman who was in the habit of telling her youthful customers that though she bought her stock of apples at the rate of three for a penny, she was in a position to retail them at the rate of four a penny, owing to the large scale upon which she conducted her purchases.

Perhaps the old dame repeated this assertion so frequently, and with such "damnable iteration," as the poet said, that at last she came to believe it, just as Beecham believes—as we have given him possible credit for believing—that the pills manufactured at his place are worth twenty-one shillings a box.

Working out the foregoing figures, according to our unsophisticated mind, Beecham ought to be a millionaire to be able to stand against his continuous heavy losses, instead of being a millionaire, as we are told, through the sale of his pills.

	£	s.	d.
Value of box of pills (according to Beecham)	1	1	0
Amount received, say	0	0	9
Actual loss per box (according to Cocker)	£1	0	3

We do not profess to set conundrums in our pages, but we will give a guinea (box of Beecham's pills) to anyone who can produce a greater puzzle, a more complete paradox than this. The Gordian knot is too tightly drawn for us to attempt to undo it. We must, therefore, try to solve the mystery by calling in the aid of a skilled analyst, who will be able to tell us what these precious pellets—these "pearls of great value," "more precious than gold or silver," as Beecham modestly styles them in the printed circular accompanying each box—are composed of.

Analytical Laboratory,
Paddington, W.,
December 21st.

Dear Sir,—On December 15th I received from you a box of ,, Beecham's Patent Pills." The box was securely fastened with the unbroken label of the Inland Revenue Office.

I have now made a careful chemical and microscopical examination of the pills.

The mass of the pill material consists of ground ginger.

The active ingredient of the pill is aloes.

In my opinion the pills consist solely of aloes and ginger mixed up with soap.

    Yours faithfully,
     ALFRED W. STOKES, F.C.S., F.I.C.,
      Public Analyst.

Goodness gracious! gracious goodness! as our ancient apple-woman ejaculated, one dark November night, when a mischievous urchin discharged a specially spiteful cracker under her humble stall. We have, it seems, been discussing the "worth a guinea a box" question upon wrong data; and, seeing that Beecham's assertion about the value of his pills is completely upset by Mr. Stokes' analysis, we are bound to admit one fact—whatever else may be disproved—namely, that Beecham is not such a loser after all.

Further investigations demonstrate that the proximate proportions of the three ingredients named in Mr. Stokes' report are as follows:—Soap, 1 part; ginger, 2 parts; aloes, 2 parts. What a revelation to be made on St. Thomas's Day—not Thomas Beecham's by any means! Well, if Beecham's scriptural namesake had had to do with Beecham's pills, considerable latitude for scepticism should have been allowed to him.

Soap, ginger, aloes! According to Beecham's assertive advertisements, some thirty or so of little pills, composed of three materials of about the cheapest possible character, are worth a guinea a box! In other words, even averaging thirty-six in a box—we are not particular as to one or two more or less, so long as we are not expected to swallow either them or the statement—each pill may be calculated at 7d.

We do not wish to be too critical, but we cannot help wondering in what part of the world Beecham buys his soap, his ginger, and his aloes to bring his pills up to this boasted value.

Tons of soap, tons of ginger, and tons of aloes made into tons of pills (otherwise "pearls of great value") ought to bring in something "more precious than gold or silver" of allegory, namely, "brass," as our Lancashire friends would say; for, as we reckon the results of such manufacture, if it were not for the patent medicine stamp, Beecham's pills would not cost even one penny to make a boxful. But the Inland Revenue Stamp comes in useful, as it enables Beecham to call his nostrum "Patent Pills," and thus convey to customers the mysterious idea that they have some remarkable special properties.

And so they ought to have if there is sufficient foundation for the high praises bestowed upon them by Beecham, in his circular, wrapped around every box of pills. They are suited to "females of all ages," says Beecham. This is a good sweep of the net, when it is considered that the feminine outnumbers the masculine portion of the community; but Beecham scorns to do things by halves, and consequently we were not surprised at finding the pills recommended, at page 5 of his circular, for "every class of disorders that afflict *all ages* and *both sexes,* from youth to old age." Here is a still wider sweep of the net, seeing that it takes in every human creature! Whatever else might be said of Beecham, he cannot be accused of leaving too many chances for other patent medicine vendors.

After the statement just quoted, others seem, by comparison, mild. Still, we may refer to some. The pills "may be given to an infant, or to the aged and infirm with perfect safety"; they "give tone and energy to the muscles, and invigorate the whole nervous system"; they "produce sound and refreshing sleep"; they are "gude for sair een," as a Scotchman would say, or, as Beecham puts it, "the eyesight is strengthened beyond conception"; they destroy "the seeds or *sympton*s"* (this is Beecham's spelling, not ours) "of direful diseases."

In all the affections he names large doses are recommended, and Beecham shows a very decided tendency towards liberality in this respect; "he giveth with a free hand," as becomes a philanthropist who is perpetually distributing guineas in return for pence. If Beecham sold his pills at the value he puts on

---

* *Itching* of the breast or head is a "*sympton*" which Beecham mentions elsewhere.

them, a guinea a box, taking his pills would be like swallowing money. For persons "labouring under the influence of drink" he advises six or eight pills as a dose. Rough on those who overstep the bounds of moderation at any festive season, some of our readers may think; but then, if, as Shakespeare wrote, men "put an enemy in their mouths to steal away their brains," they must bear the consequences, and they are entitled to scant sympathy even should they take the entire contents of the box.

The quotations we have given are sufficient specimens of Beecham's modesty. Indeed, he is sometimes so carried away by his feelings that he tries to convey to his customers that he is a qualified doctor. But he has not the slightest right to do this. A search through the Medical Register failed to discover the name; and a query addressed to the secretary of the Pharmaceutical Society elicited the answer that the name of Thomas Beecham does not appear either in the register of the members of that society or in that of the Chemists and Druggists of Great Britain. Yet this quack writes in the following misleading style:—

"It falls to the lot of very few *practitioners* to go through the experience that I have had in this class of diseases," he writes in his pamphlet, "though for many years I have not treated on this subject here, neither was it my intention to do so, but, as everyone has a duty to perform, I feel that I shall not be doing my duty if I any longer withhold that advice and information which thousands of the human family stand in need of." Briefly, the advice, given by a man whose qualifications to give professional advice may be summed up in the short word *nil*, is to persevere in taking his triple concoction of soap, ginger, and common aloes—six pills as a dose—while all the information this self-assumed medical authority (!) vouchsafes is, that "BEECHAM'S PILLS will be found to be pearls of great value," and that "they will, as sure as water quenches thirst, search out," and thus cure the most terrible diseases to which human flesh is subject. Remarkable information, truly; but, in his anxiety to get people to swallow half a box of pills daily he has forgotten to tell us which of these three ingredients in his pills is the highly vaunted *specific?* Is it the soap, or the ginger, or the common aloes? The last-named, which is merely an ordinary purgative, is the only really active ingredient of Beecham's Pills, just as it is of the Sequah Prairie Flower Mixture, of Holloway's Pills, of Mother Seigel's Syrup, of William's Pink Pills for Pale People, and many other nostrums. The ancients enumerated four elements, viz.: air, earth, water, and fire. If patent medicines had existed in their days, aloes might possibly have been included as a fifth, so frequently is it to be found in quack preparations.

Chapter XIII.

The Alofas (*All-a-Farce*) Safe Remedies, and the School of *Safe* Medicine.

"There is no composition in these news
That gives them credit."
—*Shakespeare.*

Some of the quack nostrums which have been exposed in our columns have been chiefly remarkable for the cool assertions of the makers that they have discovered a remedy of priceless value,—aloes, it may be, or soap, or saltpetre, or some equally cheap and common substance,—before which all human ills will vanish as rapidly as dew before the morning sun. In the present instance we have to deal with a so-called company, which parades a whole string of preparations—almost as long as the list of diseases which they "never fail to cure"—at a shop in New Oxford Street.

Certainly, some little modesty, an article as rare as truth amongst quacks, is exhibited in the appropriately green-covered pamphlet which is freely distributed by the Alofas Company, for they say: "At a time when so many drugs and patent medicines compete for public favour, it may seem to need some apology for adding to their number." But Modesty, blushing to find herself reluctantly pressed into the service of a patent medicine man, soon disappears, the scribbler resumes his wonted impudence, and writes as follows:—" The excuse is to be found in the fact that modern medical science, in spite of its varied resources and the agents now at its command"—patent medicines thrown in, we suppose—" still too often finds itself unable to cope with the various diseases which attack humanity." So that it would appear that the Alofas Company, *alias* a man named " Younger," must have been sent into the world with a special mission to cure where medical science fails, or that, to adopt a homelier phrase, " fools step in " where science carefully treads. However, we must not be altogether ungrateful; the writer of the pamphlet—for short, let us say, Younger—gives his readers a faint glimpse of the curative paradise which bounteous Nature has, in some mysterious manner, unfolded to his wondering gaze; he tells them that rare exotics have a wide range of curative power, especially in the more difficult and complicated disorders, and that " it has been the object of the Alofas Company (with the assistance of a gentleman who, both in the Old World and the New, has made a life-long study of the botanical productions of this and other countries) to make that happy combination of British and exotic plants, herbs, and flowers, to which the marvellous success of the Alofas Company's remedies is to be attributed." Of course, it goes almost without saying that these choice herbal and floral tributes of Nature to Younger's penetrating knowledge are described as absolutely harmless; why, we will explain shortly, and we may even now concede that this statement has some amount of truth about it. But, the same reason compels us (as we shall also presently show) to flatly contradict Younger's assertion that "these preparations are in no sense quack medicines, but, on the contrary, are curative agents of great and tried value."

The list of Alofas medicines is of alarming dimensions, and ranges from Anti-germ smelling-bottle, 2s. 9d. (reminding one of the pills against earthquakes sold by a charlatan many years ago), down to a special female tincture. They cure everything, so Younger says; for instance, the Alofas tincture "cures all chest and throat diseases, including consumption, bronchitis, and pleurisy"; the Alofas powder "cures all wasting diseases, night sweats, debility, brain fag, &c."; the Pills promptly settle "all liver and bowel disorders"; the Stomachic removes "all kidney and heart troubles"; the Safe Remedy for Corpulence would bring a modern Daniel Lambert down to a Derby jockey's riding weight in a jiffey, and so on through the list. "You pay your money, and take your choice," as the showman says, of the sixteen Alofas compositions, and Younger guarantees a safe, a pleasant, and a rapid cure. What more could a sensible man require, except that some one, whose undertaking would be worth accepting, would guarantee Younger's veracity.

Our readers would, we are convinced, not be so unreasonable as to expect us to invest good money, and to waste time which might be put to better use, in purchasing and examining all the sixteen Alofas medicines. In the embarrassment of such an amount of remedial wealth, we felt that the best course to adopt would be to sample; after which our readers could follow the advice which Batty used to advertise concerning his condiments, "If you like the pickles, try the sauce."

Consequently we bought a small bottle of the first thing on the list of the "safe herbal specialities"; we had some difficulty in limiting the extent of our purchase, as the youth who was at the time of our call apparently in sole charge of this widely-ranging set of specialities, most energetically prescribed a larger sized bottle, to be taken in conjunction with several of the other remedies. But we were proof to the persuasive efforts of this junior—he might have been a Younger as well, judging by his intense anxiety that we should take the greater part of the visible stock away with us; and after ascertaining that if we wished we could consult " Dr." Younger (" not an M.D. of London," as the boy told us) on any evening, we retired, *minus* 13½d., and *plus* a bottle of Alofas tincture, described by Younger, in his pamphlet, as "a safe cure for *consumption*, bronchitis, pleurisy, and all throat and chest diseases"; this being as gross a misrepresentation as was ever put on paper.

A tincture is defined in chemical treatises as "a solution of a medicinal substance in alcohol." Now, there was not a drop of alcohol in the whole bottleful of fluid; so Professor Wanklyn, who analysed the stuff for us, has reported: and the only "medicinal substances" were glycerine coloured with treacle, of which "rare exotic" productions Younger had lavishly put 15 per cent. (more than a teaspoonful to the ounce) into the contents. A somewhat disagreeable bitter flavour, evidently introduced to make the stuff taste more "phisicky," as credulous old women would say, struggled with the sickly sweet treacle for predominance. A very close imitation of this "sure cure for consumption" might be made as follows:—
Put nearly three and a half ounces of water, previously passed over quassia or chiretta chips (both inexpensive, for the ingredients used by patent medicine makers are only rare and costly in imagination and misrepresentation, not in fact) to give a flavour, then add more than half an ounce of glycerine and treacle, cork it, shake it up well (the youthful prescriber artfully did this, as a final precaution, we noticed, when he sold us our treacly treasure), clap on a Government three-halfpenny stamp, and get the first person who believes that this composition is "a safe cure for consumption and all chest and throat diseases" to

give you good silver coin of the realm for what would not cost two farthings to make.

From what we happen to know, there is full reason for the belief that the other preparations of the *All-a-Farce* Company are of the same absolutely worthless character as the tincture.

Perhaps our readers may be interested to learn who this D. Younger, sometimes called Dr. Younger, is. There is an institution at 21, Stepney Green, in the East End of London, styled by its promoters the Magnetic and Botanic School of Safe Medicine The leading men connected with it are, Charles Gapp, until a few years ago a coal-dealer's clerk, but now carrying on a quasi-medical practice in the East End of London, and describing himself on his letter paper and in print, as an M.D., though his only claim to that title is a bogus diploma which he obtained from the United States without even leaving London; a "reverend" Verryman Trimmings, whose Ph.D. and M.A. have a similar origin to Gapp's M.D. diploma; and "Dr." Younger, also possessor of a diploma like Gapp's, the manager of the *All-a-Farce* Company, F.A.I. of Paris—whatever that may mean—and president of the institution at Stepney; together with one Maguire. From a prospectus, bearing a recent date, we gather that the Magnetic and Botanic School of Safe Medicine (practically the quartet we have named) undertake "to impart knowledge on safe medicine" by a sort of postal tuition. This tuition is described as being "a unique system, plain to the intellectual grasp, and thorough in its mode of imparting knowledge," and "specially adapted for persons engaged in business during the day"; the postally "posted" being entitled in due time, and doubtless after a due payment, unspecified, to receive a diploma, which is "a very handsome production and well worth a mental effort" (and a good, round sum to the postal professors) "to secure." As glycerine and treacle are regarded by the president as a "safe cure for consumption, pleurisy, bronchitis, and all chest and throat affections," the *materia medica* taught at the Stepney establishment must be of the same class as that referred to in the following advertisement which appeared some months ago in a provincial daily paper:—"Wanted, an agent to undertake the sale of a patent medicine. The proprietor guarantees that it will be profitable to the *undertaker*." "Which nobody can deny," as the old song says; but what we do deny, most emphatically, is the wisdom of the authorities in allowing Younger & Co. to issue broadcast diplomas certifying medical efficiency, and thus calculated to deceive unwary and ignorant people, although their only value, according to the printed admissions of Younger & Co., is that they are "very handsome productions," or, as we should put it, *imposing* documents.

Chapter XIV.

The Ignorance of Quacks; the Blindness of their Dupes.

"And yet with canting, sleight, and cheat,
'Twill serve their turn to do the feat."— *Hudibras.*

THE ignorance of quacks is equalled only by the impudence of their assertions. Their sole aim and concern is

"Not for the sickly patient's sake.
Nor what to give, but what to TAKE."

Provided that they can succeed in this last-named respect, their object is accomplished.

In the course of these articles their ignorance of all medical matters is proved to be appalling. One quack will pretend that there is only one disease, for the simple reason apparently, that he has only one nostrum to offer; another will assert that his preparation is imbued with electric properties, though a fluid of exactly corresponding character can be drawn from any water-tap; and a third will unblushingly proclaim through every paper in the land that he is the sole possessor of a rare and hitherto unknown exotic, which turns out on analysis to be nothing more than a very indifferent specimen of common aloes.

Alcohol has been described by a well-known temperance advocate, Sir B. W. Richardson, F.R.S., as the "devil in solution." Quack medicines would appear to be the devil both in solution, and in a solid form, judging by the number of lies—of which his Satanic Majesty is credited with the paternity—uttered for the purpose of promoting their sale.

The credulous customers of the quack medicine men are generally, too, quite as ready to swallow both the falsehoods and the physic as the unscrupulous vendors are to dispose of the latter. This form of blind faith is not confined, either, to those whose want of education might be urged as some excuse for them; for it is not unfrequently to be found amongst persons of a higher social status. In such people, the faith in some particular nostrum amounts to a sort of fetish-worship, the foundation of which is based on "cussedness," as our American friends would say, on a dogged determination to hold an opinion, however devoid of reason it may be. Truly, as the poet whom we have quoted at the head of this chapter wrote :—

"Obstinacy's ne'er so stiff,
As when 'tis in a wrong belief."

Speaking just now of the ignorance of quacks, we are naturally reminded of many anecdotes concerning their general as well as their medical ignorance; but we have only room for two of these.

The first is of a man who was tried many years ago at the Old Bailey for illegal medical practice, which had culminated (as the administration of quack medicines must often do) in the death of a sick person with whose case the charlatan had tampered.

"What are you?" asked the prosecuting counsel; "A Surgin," was the reply. "How do you spell it?" "Z-u-r-g-i-n." "*How?* excuse me; I did not quite hear what you said." "C-u-r-g-i-n." (Laughter in Court.) "You sometimes bleed people, do you not?" "Yes, of course I do." "Where do you bleed them?" "In the temporal artery." "Where is the temporal artery?" "Why, in the arm, to be sure; you don't think you can catch me, do you? I am too good a dentist not to know where the temporal artery is." (Renewed and prolonged laughter.)

The other anecdote is of an American quack. A New York subscriber to HEALTH NEWS has informed us that, some time back, there was picked up in one of the streets of that city a pocket-book, evidently the property of one of the irregular practitioners with whom New York abounds. The loser had an eye to business and dollars, as evidenced by the following *verbatim* extracts:—

"Kase 230, *Mary An Perkins*, Bisnes, washerwoman. Siknes in her hed. Fisik, sum blue pills and a soaperifik, age 52. Ped me one dollar, one kuarter dollar pece bad. Mind get good kuarter dollar pece, and make her take mo fisik.

"Kase 231, *Tummus Krinks*. Bisnes, Irishman, and lives with Paddy Molony whot keeps a dray. Siknes, dig in the ribs and two black eyes. Fisik, to drink my mixter twise a day of sasiperily beer and jollop and fish-ile with asifedity to make it taste mo fisiky. Rubed his face with kart greese linnyment, aged 39 years old. He drinked the mixter and wuddn't pay me bekase it tasted nasty, but the mixter will make him feel sorry, I reckon.

"Kase 232, *Old Muther Boggs*. Ain't got no bisnes, but plenty of money. Siknes awl a humbug. Guv her sum of my celleberated 'Dipseflorikon' wich she said drinked like coled tea, wich it was too. Must put sumthink in it to make her feel real sik and bad. The old wummun has got the fidgitts."

Chapter XV.

The Gold Cure for Drunkenness.

"Men often swallow falsities for truths."—*Sir Thomas Browne.*

IF the learned writer from whom we have taken this quotation (worthy to rank with John Locke as foremost of philosophical physicians) could get an insight into the quackery of the present day, he would find abundant opportunity of verifying his axiom.

Indeed, the field is so rich in illustrative facts that one feels an *embarras de richesses* in selecting one for the purpose. As we write these lines, and glance at the papers lying upon our library table, we catch sight of a boldly displayed advertisement, printed throughout in capital letters, on the leader page of one of the principal London daily newspapers. "Drunkenness permanently cured in three weeks," says this announcement, "by the double chloride of gold treatment as prepared and given with such marvellous results by G. H. McMichael, M.D., of Niagara Falls, U.S.A., in his various institutes in America." Further on, we are told that at the British Gold Cure Institute, located for the time being in a street at the West End, dipsomania or hereditary drunkenness is permanently and easily cured. Hereditary drunkenness! This is an ingenious way of helping the drunkard out of his difficulty and disgrace; just as, in the case of gout, no one was ever known to admit that his disease was of his own making—the blame is invariably fixed on a grandfather or some equally convenient ancestor—so we shall have every sot excusing his delinquencies with the remark, "Can't help it, it's (hic!) hereditary, don't you know?"

Now this alleged British Gold cure turns out to be what might with greater truth be designated the American Brass humbug. Its home is at Dwight, in Illinois, and one strong reason for its finding its way into this country, apart from John Bull's proverbial gullibility, is that it has been, to use an Americanism, pretty nearly "bust up" in the United States.

Many of our readers may have noticed in small shops in country districts base coins securely fixed to the counter, by way of reminding any utterer of dishonest money who may enter the place that the tradesman is on his guard against trickery and fraud. Following this example, we will just drive a nail into one lie concerning the bichloride of gold treatment. An eminent analyst, who has examined this preparation, reports that it contains neither gold nor chlorides, but that its composition is as follows:—Water, 61·31 per cent.; sugar, 6 per cent.; a small quantity of lime-salts; and 25·55 per cent. of Alcohol. *Neither gold, nor chlorides* of any metal! Who, after this startling revelation, would dare to attempt a contradiction of Sir Thomas Browne's *dictum*, practically applied in this instance, "Men often swallow falsities for truths."

But, though the poor dupes of the American Brass curers of drunkenness (always hereditary, be it remembered, according to their veracious statements) get no gold in their physic bottles—the price charged for those which were analysed was nine dollars (36s.) for two bottles, the lowest number which a purchaser could be supplied with—they get, instead of the expected gold, a wholly unexpected ingredient, in the form of alcohol, to the extraordinary extent, considering what the medicine is given for, of considerably more than one-fourth of the entire quantity of fluid; thus making it double to the strength of champagne, and equal to that of port or sherry.

If, however, the patient gets no gold, the same cannot be said of the conductors of the Institute. In reply to a gentleman who wrote, asking for the terms upon which the cure would be supplied, the secretary, Mr. Gerard D. Apthorp, informed him by an officially written letter that the charge was fifteen guineas for three weeks, with five guineas for each additional week. Fifteen guineas were, really, the lowest amount that could be accepted for such a priceless boon; and the continuance of the treatment after three weeks, at the modest figure of five guineas for every seven days, would be dependent on the judgment of the doctor—subject, doubtless, also to some extent to the question whether the patient, having swallowed so little gold and so much alcohol in the three weeks' course, had any more of the former indispensable in his pocket. "We do not send out medicines, as we find that to be successful," the secretary somewhat vaguely wrote, "patients must come to the Institute"—where they have to submit to periodical hypodermic injections of narcotic poisons, in addition to quaffing their *golden* grog.

At a meeting of the London Society for the Study of Inebriety, presided over by Dr. Norman Kerr, who has done much to expose the Gold Cure humbug, the chairman said that an attempt, fortunately frustrated, had been made to float a syndicate to raise £150,000 in this country to purchase the right of using the Gold Cure; £110,000 in hard cash and £40,000 in shares, being all the proposing vendors asked! Such unparalleled philanthropy ought not surely to pass unnoticed.

Some of the rules of the Institute are too interesting to escape notice, too, if we read between the lines. The first two rules refer, as we need scarcely observe, to fees and prompt payment, patients being delicately requested "to arrange all financial matters with the Secretary-Treasurer on their arrival."

Rule 4 lays down that "the remedy for internal use is compounded to meet individual needs"—can this have reference to the relative proportion of alcohol?—"and loaning or exchanging is not permitted." We could understand that "loaning" would be in brisk requisition if the bottles actually contained —what they are presumed to contain—gold; but what gentleman or lady, except a drunkard, hereditary or otherwise, would dream of "loaning or exchanging" a bottle of liquor?

Rule 5 there is little to find fault with, save that it does not go far enough. "Bathing is essential, and patients are required to bathe at least twice each week." Drunkards are commonly more chary of using soap and water externally than they are of using brandy or whisky and water internally; so that there is as much need of advice of this kind as there was in the subject of the following anecdote. A particularly dirty-looking individual, dirty enough to have been a hermit had he lived centuries ago, presented himself at the consulting rooms of a West End physician, to seek advice concerning a troublesome cutaneous eruption, entirely due to his uncleanly habits. The physician, after a little preliminary conversation, commenced to advise the patient, who, of course, was all attention. "I should recommend you to get forty gallons of boiling hot water into your bedroom when you go home, and to place it in a large tub; next, having divested yourself of your clothes, to immerse your body in the hot water; then, taking a piece of soap in your hands, to rub it well until you have formed a good lather, which you should carefully proceed to apply all over your body, removing it by the aid of hot water. Afterwards you should rub the surface of your body with a dry rough towel." "It seems to me." grunted the indignant patient, scarcely able to restrain his rising passion, "that you are telling me to have a bath." "It does appear like it," blandly observed the physician, as he swept the fee with a clinking sound from the table into a small drawer, "and by the bye, I would recommend you to persevere with the prescription."

Rule 7 reads as if it had reference to the alcohol contained in the golden solution: "The physicians earnestly entreat patients to avoid saloons and bar-rooms, and to use only what is prescribed at the office." Why, certainly! If a man could not content himself with a potion as strong as port or sherry, and twice as powerful as champagne, he would be a most unreasonable individual.

At the meeting of the Society to which we have already referred, Dr. Usher, of Melbourne, Australia, described a visit he had made to the Gold Cure Institute at Dwight. The proprietor, "Dr." Keeley, who was very uneasy while conversing with him, told him he had employed "the remedy" twelve years with great success. To Dr. Usher's very natural questions as to the treatment and what prescriptions he used, Keeley replied, "We will not go into that; I know it is all right. If you want to learn anything about it, the secretary-chemist will tell you." This secretary-chemist turned out to be a sort of page-boy. Dr. Usher was introduced to Dr. Blaine, " the chief of the staff"; he did not tell his audience what " the chief of the staff" was like, but our readers can perhaps arrive at a fairly accurate conclusion when we mention that " the staff consisted of unsuccessful practitioners." Dr. Usher was taken where three rows of men were being injected in the left arm with five drops each, out of a little porcelain bowl containing a pinkish solution—atropine. Many of the patients wore glasses, and they told him that they could not see three or four days after the commencement of the course of treatment, and had become almost blind. They suffered, too, from giddiness. One patient had been at the institution nine weeks, and was afraid to leave because three of his "pals" who had left ten days before had got drunk and had to come back. So much for the permanence of the Gold Cure! Another patient, who really seemed in some respects more intelligent than the rest, candidly imparted the imformation that an aunt had promised him an annuity if he stayed there two or three months, so that he " wanted to see the time out."

Incidentally, Dr. Usher throws an amusing light on the manufacture of testimonials. Three leading physicians from New York, Boston, and Philadelphia, representing three different societies in those cities, had visited the Institute, at various times, remaining half a day or so. Some six weeks afterwards, Keeley (the sly rascal!) issued a number of circulars which set forth these three gentlemen as remarkable instances of successful cure. One of the special features of the Gold Cure Institutes is that they have amongst the members a "Bichloride of Gold Club," including two classes termed, respectively, "graduates" and "undergraduates." Could anything beat this as a specimen of exquisite fooling?

It is too often the practice in England to severely punish small offenders, while great ones go scot-free.

In the daily paper from which the advertisement of the Gold Cure Institute was taken, we read a policecourt report of a miserable old man who was sent to prison for trying to pass a counterfeit shilling in payment for a scrap of bacon. Decrepit dolt! If, instead of resorting to what may be termed the silver cure for starvation, and tendering a coin deficient in value by only a few pence, he had hired apartments in a fashionable quarter and passed off a compound of alcohol, water, and sugar as containing gold, he might have "gone the whole hog" in place of a beggarly bit of bacon, and have made thousands upon thousands of golden sovereigns in a few months.

Who is responsible for the inaction of the police, who is to blame for the inertness of the medical corporate bodies, in such gross instances of quackery and imposture as we have exposed, and shall continue to expose, in the columns of HEALTH NEWS?

END OF VOLUME I.

Savoy House, 115, *Strand*, W.C.

The following are some of the Publications of this firm:

HEALTH NEWS (Illustrated), Established in 1887. The best, cheapest, and most widely circulated Health Journal. 3d. Monthly, post free for 4 stamps. Subscription for Twelve Months, commencing at any date, 4s., post free to any address in the kingdom.

EXPOSURES OF QUACKERY; containing a Series of interesting Articles upon, and Analyses of, the Principal Patent Quack Medicines. In two volumes, 1s. each, post free for 14 stamps.

STAMMERING, STUTTERING, AND OTHER SPEECH AFFECTIONS. 1s., post free for 14 stamps.

HAY FEVER, HAY ASTHMA, OR SUMMER CATARRH 1s., post free for 14 stamps.

DEAFNESS, NOISES IN THE EARS, &c. Their Causes and Treatment. 6d., post free for 7 stamps.

CHOLERA: Its Nature, Causes, and Cure. 3d., post free for 4 stamps.

ADVICE TO A WIFE, ABOUT HERSELF AND BABY. 1s., post free for 14 stamps.

FIRST TEN YEARS OF A DOCTOR'S LIFE. 1s., post free for 14 stamps.

GOUT: ITS NERVOUS ORIGIN. 1s., post free for 14 stamps.

INFLUENZA: Its Nature, Symptoms, and Treatment. Second Edition. 1d., post free for 3 Half-penny stamps.

HYDROPHOBIA AND DISTEMPER MADNESS. A Plea for the Canine Race. 1d., post free 3 Half-penny stamps.

DIABETES AND OTHER URINARY AFFECTIONS. 2s. 6d., post free for 2s. 9d.

THE
ANTI-ADULTERATION
ASSOCIATION,
(REGISTERED 1895)

61 & 62, CHANCERY LANE, LONDON, W.C.

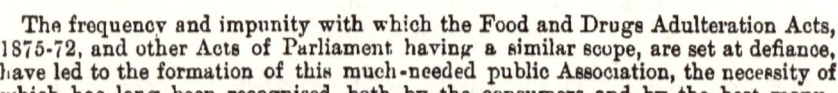

The frequency and impunity with which the Food and Drugs Adulteration Acts, 1875-72, and other Acts of Parliament having a similar scope, are set at defiance, have led to the formation of this much-needed public Association, the necessity of which has long been recognised, both by the consumers and by the best manufacturers and retailers.

The following are amongst the desirable objects which the Association has in view:—

1.—To put more regularly into force the provisions of the existing Acts, and to prevent their becoming practically inoperative, through the energetic greed of dishonest traders, the apathy or helplessness of consumers, or the inadequate punishment dealt out to offenders in the majority of cases where offences against the Acts have been proved.

2.—To obtain by Statute such modifications of the Acts—at present of too permissive a character, and affording too many loopholes of escape for wrong-doers—as are necessary in the interests alike of the public, and of honest, fair-dealing manufacturers and retailers.

3.—To protect consumers by publishing, from time to time, reports and analyses, together with other useful information concerning foods, beverages, drugs, &c., and their adulteration or purity.

4.—To protect and encourage honest dealers and genuine manufacturers by the issue of official certificates of Purity and Excellence, under certain regulations.

The annual subscription of members is fixed at the small sum of half-a-guinea, entitling the member to HEALTH NEWS, the organ of the Association, sent monthly, post free, and to admission to all lectures and meetings. A donation of £3. 3s., or upwards, constitutes life-membership. Members will also be entitled to have analyses made at reduced terms.

Further particulars, concerning membership, terms for analyses, and regulations for granting certificates can be obtained of the Hon. Secretary. Subscriptions should be made payable by postal order and cheque, crossed "Cheque Bank."

N.B.—Analyses of all kinds conducted in the Laboratory, for non-members, at very reasonable charges.

GOODALL'S HOUSEHOLD SPECIALITIES.

YORKSHIRE RELISH

THE MOST DELICIOUS SAUCE IN THE WORLD

This cheap and excellent Sauce makes the plainest viands palatable, and the daintiest dishes more delicious. The most cultivated culinary connoisseurs have awarded the palm to the YORKSHIRE RELISH on the ground that neither its strength nor its piquancy is overpowering, and that its invigorating zest by no means impairs the normal flavour of the dishes to which it is added. Employed either *au naturel* as a fillip to chops, steaks, game, or cold meats, or used in combination by a skilful cook in concocting soups, stews, ragouts, curries, or gravies for fish and made dishes.

The only cheap and good Sauce. Beware of Imitations.

Sold in Bottles, 6d., 1/-, and 2/- each.

GOODALL'S JELLY SQUARES

Make delicious and nutritious Jellies in a few minutes and at little expense. The Squares are complete in themselves for making Cherry, Champagne, Brandy, Lemon, Orange, Raspberry, Strawberry, Vanilla, Pineapple, Black Currant, Red Currant, Almond, Plain, Port and Sherry Wine Jellies, and are sold in **boxes containing half-pints, pints, and quarts, at 3d., 6d. and 1s. each.**

Sold by Grocers, Chemists, Oilmen, &c.

PROPRIETORS—

GOODALL, BACKHOUSE & Co.,
WHITE HORSE STREET, LEEDS.

AWARDS:

GOLD MEDAL, International Health Exhibition, London, 1894.
FIRST ORDER of MERIT & MEDAL (Highest Award), Adelaide, 1887.
HIGHEST AWARD. Medical and Sanitary Exhibition, London, 188½.
FIRST ORDER of MERIT & MEDAL, Melbourne, 1882.

BENGER'S FOOD

(REGISTERED)

For INFANTS, INVALIDS, and the AGED.

THIS delicious and highly nutritive Food was awarded the GOLD MEDAL at the International Health Exhibition, London, and has since received a High Award at every Exhibition at which it has been shown.

BENGER'S FOOD is well-known to leading medical men, and is recommended by them.

The LANCET of March 25, 1882 says:
"MR. BENGER's admirable preparation."

The BRITISH MEDICAL JOURNAL August 25, 1883, says:
"BENGER'S FOOD has, by its excellence, established a reputation of its own."

The LONDON MEDICAL RECORD March 5, 1882, says:
"It is palatable and excellent in every way. It is taken readily both by adults and children; we have given it in very many cases with the most marked benefit, patients frequently retaining it after every other food has been rejected. For children who throw up their food in curdled masses it is invaluable."

FROM AN EMINENT SURGEON.
"After lengthened experience of Foods, both at home and in India, I consider BENGER'S FOOD incomparably superior to any I have ever prescribed."

EXTRACTS FROM PRIVATE LETTERS.
"The infant was delicate, and our medical adviser ordered your Food. The result in a short time was wonderful, the little fellow grew strong and fat, and is now in a thriving condition — infact, ' the flower of the flock.'"

', I do not think I should be doing my duty if I did not speak up for ' BENGER'S FOOD.' It has simply been the means of bringing my baby boy back to life. I enclose his photo that you may see what a bonny boy he is."

"My last little boy was fed entirely upon it from birth, and a healthier child it would be difficult to find. My wife sounds the praise of BENGER'S FOOD everywhere."

"I have very much pleasure in stating that when all other Foods failed, yours was recommended to us by Dr. ——, and has been the means of saving my dear little daughter's life."

BENGER'S FOOD is sold in Tins by Chemists. &c., everywhere.

"Your Preparation is certainly the best I have ever examined."—H. C. BARTLETT, PH D., F.C.S.

NINE First-Class Exhibition Awards.

SCOTT'S
MIDLOTHIAN
OATFLOUR

UNEQUALLED as the most wholesome and nutritious Food for INFANTS, YOUNG CHILDREN, and INVALIDS.

It contains in concentrated form those flesh and bone properties peculiar to Oats, and is ENTIRELY FREE FROM HUSK.

Samples free to Members of the Medical Profession on application to the—

SOLE MAKERS:

A. & R. SCOTT, LD., GLASGOW & LONDON.

CADBURY'S
COCOA

Absolutely Pure, therefore Best.

GENUINE COCOA.

The public are warned against chemically prepared dark liquor cocoas claiming to be "pure," but in reality prepared with a considerable per centage of alkali; this can be detected by the unpleasant smell when a tin is first opened.

Cadbury's Cocoa, on the other hand, is guaranteed to be absolutely pure, and can be safely and beneficially taken as an article of daily diet AT ALL TIMES AND SEASONS.

THE LANCET SAYS:
"Cadbury's Cocoa represents the Standard of highest purity at present attainable."

"TO QUACK OF UNIVERSAL CURES."—HUDIBRAS.

EXPOSURES OF QUACKERY:

BEING A SERIES OF ARTICLES UPON, AND ANALYSES OF,

NUMEROUS PATENT MEDICINES.

BY THE
Editor of "HEALTH NEWS."

VOLUME II.

LONDON:
THE SAVOY PRESS, LTD., SAVOY HOUSE, 115, STRAND. W.C.

PRICE ONE SHILLING.

(COPYRIGHT.)

'HEALTH NEWS'

(ILLUSTRATED),

Monthly, 3d., Post Free for 4 Stamps.

THE BEST, CHEAPEST, & MOST WIDELY-CIRCULATED HEALTH JOURNAL.

THE OFFICIAL ORGAN OF THE ANTI-ADULTERATION ASSOCIATION.

The Number for January, 1896, began a New Volume, but subscriptions can commence from any date. Post free for 12 months, 4s.

Devoted to the consideration of Public and Individual Hygiene, House Construction, Dietetics, Foods, Beverages, Adulterations, Health Resorts and Mineral Springs, Domestic Sanitation and Regimen for Invalids, Sanitary Inventions, Literature, &c.

Excellent Advertising Medium for all announcements intended to reach the well-to-do Classes. Rates Moderate. Quack advertisements are rigorously excluded.

THE SAVOY PRESS, Ltd., SAVOY HOUSE, 115, STRAND, LONDON.

PRINTING and PUBLISHING

THE SAVOY PRESS, Ltd.,

SAVOY HOUSE, 115, STRAND, W.C

Are prepared to undertake the

PRINTING AND PUBLISHING OF WORKS IN EVERY CLASS OF LITERATURE AT MODERATE CHARGES.

MSS. read and Confidentially advised upon.

ESTIMATES FREE.

EXPOSURES OF QUACKERY.

VOL. I.

Exposures of Quackery.

The following are a few of the hundreds of favourable Press Notices which have appeared.

The **Times** says: "A cordial recognition is due to the great public services rendered by the republication of these articles."

The **British Medical Journal** says:—"We hope that all our readers will make it their business to acquaint themselves with these exposures."

The **Medical Press** says: "We hope that the good seed of knowledge sown broadcast by this courageous editor will fructify to some future purpose in the minds of the community."

The **Provincial Medical Journal** says:—"Every medical man should make himself acquainted with these articles on patent medicines, and make them generally known also amongst his patients."

The **Saturday Review** says:—"The editor has issued a series of articles exposing the pretensions of popular patent medicines. He has furnished thinking people with weighty reasons."

Public Opinion says:—"A perusal of these articles by our able contemporary is at once interesting and instructive, as all the well-known patent medicines are criticised in its pages. The articles are brightly written."

In two vols., 2s. To be had direct from the Savoy Press, Ltd., 115, Strand, London, post free for 2s. 3d.

[For Contents of Vol. 1, see the page facing Preface.]

"TO QUACK OF UNIVERSAL CURES."—HUDIBRAS.

EXPOSURES OF QUACKERY:

BEING A SERIES OF ARTICLES UPON, AND ANALYSES OF,

NUMEROUS PATENT MEDICINES.

BY THE
Editor of "HEALTH NEWS."

VOLUME II.

LONDON:
SAVOY PRESS, LTD., SAVOY HOUSE, 115, STRAND. W.C.

PRICE ONE SHILLING.

(COPYRIGHT.)

CONTENTS OF VOL. I.

CHAP. 1. Patent Medicines; Patent Medicine Law; Mattei's "Electricities."—CHAP. 2. Clarke's Blood Mixture.—CHAP. 3. "Protected by Government Stamp"; Chlorodyne, and other Opiates and Anodynes.—CHAP. 4. Revalenta Arabica.—CHAP. 5. The History of Patent Medicines; The Sequah "Prairie Flower" Mixture, and Oil.—CHAP. 6. Holloway's Pills and Ointment; Sequah "Prairie Flower" Mixture; Letter concerning Clarke's Blood Mixture.—CHAP. 7. Saved from the Waste Paper Basket; Correspondence concerning Holloway and Mattei.—CHAP. 8. Allen's World's Hair Restorer; Mexican Hair Renewer; Singleton's Golden Ointment for the Eyes; Rowland's Kalydor, and Gowland's Lotion for the Skin; Anna Ruppert's Skin Tonic; More about Mattei.—CHAP. 9. Quack Advertisements and Testimonials; Mother Seigel's Syrup.—CHAP. 10. Clarke's Blood Mixture; The Alleged Testimonial from the late Dr. Swaine Taylor, F.R.S.; The Obverse and the Reverse.—CHAP. 11. Quack Testimonials; Clarke's Blood Mixture; Mother Seigel's Syrup.—CHAP. 12. Beecham's Pills.—CHAP. 13. The Alofas Safe Remedies, and the School of Safe Medicine.—CHAP. 14. The Ignorance of Quacks; The Blindness of their Dupes.—CHAP. 15. The Gold Cure for Drunkenness.

PREFACE TO VOLUME II.

"The first of all gospels is this, that a Lie cannot endure for ever."—CARLYLE.

THE large demand which has followed the publication of the first volume of EXPOSURES OF QUACKERY is, of itself, fairly good proof that the articles which have appeared under that title in the columns of HEALTH NEWS have obtained an appreciative reception.

But we look beyond this for evidence that we have done public service in the efforts to accomplish a task which had never before been attempted, viz.—the production of a systematic series of reports upon, and analyses of, all the largely advertised patent medicines. There were plenty of reasons why we should not have undertaken it, seeing the amount of time, trouble, and expense involved, and the bitter hostility that it would arouse in certain quarters, through the dread of diminution of ill-gotten gains. We felt, however, that we ought not to allow ourselves to be deterred by any such considerations; we had put our hand to the plough, and it would have been an act of moral cowardice, to have turned back. Rather, we thought, on the contrary, obstacles and opposition should determine us to persevere in what has been termed by a leading contemporary, "a vigorous crusade against quackery." "Quixotic," was the remark of more than one candid critic; but we did not quite take this view. We called to mind Carlyle's emphatic words, "The first of all gospels is this, that a Lie cannot endure for ever," and we were convinced that nothing exists to which his short definition could be more appropriately applied than to quackery.

It is the meanest, the cruellest, the most objectionable, and the most dangerous form of robbery. Protracted suffering, tormenting disease, distressing anxiety to invalids and their relatives and other friends, even ultimate loss of life, through wrongly administered powerful drugs, or through delay in seeking proper, skilled medical treatment,—such serious matters as these have no weight with the quack. The highwayman of bygone times presented a pistol at his victims, with the explicit intimation, "Your money, or your life"; but, having possessed himself of the former, he rode away, leaving the robbed individuals unhurt. The modern quack utters no warning; if he did, it would very often be "Your money *and* your life," so utterly callous is he as to consequences, provided he can gratify his greed. Is it not worth while, therefore, to endeavour to check a system of plunder, of fraud, of injury to health? There cannot be anything about such a task to warrant its being characterised as Quixotic.

We have, indeed, full reason to believe that our efforts for the repression of quackery have already produced good results. In fact, we have had indirect, yet conclusive, evidence of this fact from the people most likely to know, namely, the quacks themselves. They have attacked us with abusive letters,—somewhat singularly, always anonymous,—they have threatened us with actions for libel, and they have tried to cajole us with tempting offers to fill our advertisement pages.

As to abusive letters, we accept them as complimentary, coming from such a quarter, and like the famous Jackdaw of Rheims, in the *Ingoldsby Legends*, we have been "not a penny the worse" through their display of temper, except when, in their impotent rage, the writers have omitted to affix postage stamps to their communications.

With regard to their threats of legal action, the amusing but pusillanimous character, Bob Acres in Sheridan's *Rivals*, was a paragon of pluck compared with them. Their conduct, when a legal battle seemed imminent, reminded us of the "hearts of hare" in Sir Walter Scott's poem; for their courage "oozed away" faster even than blustering Bob's in Sheridan's play, directly they found that we were prepared to substantiate the correctness of our statements, and that the analysts were equally ready to stand by their reports. We did think not long since, that a certain patent medicine proprietor would have the hardihood to go on,—nothing would have pleased us better than to have had an opportunity of exposing his pretensions in Court,—but we happened, in our reply to his solicitors, to quote the following extract from Mr. Justice Mathew's summing-up in a trial where a quack was concerned:—"A sharp, pungent, and unpleasant criticism does not constitute a libel, so long as it is honestly written." The promised writ has not yet reached HEALTH NEWS Office, and does not appear ever likely to arrive there.

With regard to repeated offers of liberal payment for the insertion of their announcements,—three or four times more than the usual rates charged by our publisher,—they became so frequent and pertinacious that we deemed it expedient to put the notice, "Quack advertisements rigorously excluded," in a prominent position on the front page of HEALTH NEWS. We may refer here, in passing, to a particularly paltry piece of revenge on the part of some patent medicine men. Finding that their abuse, their threats, and their proffered bribes, failed to produce the desired effect, they tried to bring their enormous influence, as very large advertisers, to bear on advertising agents, by urging these to secretly boycott HEALTH NEWS, and to cease recommending our journal amongst other firms doing advertising through these agents. Of course, it is exceedingly difficult to obtain evidence of this sort of thing, yet some curious proofs of its existence have come to our knowledge.

But we will turn to a more agreeable aspect of the outcome of our Exposures. We are constantly receiving letters from correspondents of all classes of society, resident in Great Britain, or abroad, expressing their cordial thanks for the denunciation of the imposture of quack nostrums. Many of these communications are from people who formerly, misled by false assertions of miraculous healing properties, pinned their faith on this or that nostrum. Others are from persons of position and education who have had exceptional opportunities of judging the immense amount of injury done to the community by patent medicines. It would occupy a volume larger than the present one, were we to print all this correspondence, and we must therefore content ourselves with quoting only a single letter. A Warwickshire rector writes to say that, although his parish is thinly populated, he has known numerous instances of quack medicines being used, "in vain, of course, by very poor people, utterly unable to afford to be swindled in such a way; and I have no doubt that every clergyman throughout the country could tell a similar tale. I feel that we all owe you many thanks for so fearlessly denouncing the imposture of quackery, and I am sure that much good must come from your efforts." Our clerical correspondent concludes by expressing the belief that the reprints of our articles "will help to stop the widely spread practice of quack medicine drugging, responsible for many deaths."

We should be wanting in gratitude if we omitted to heartily acknowledge the unstinted approbation and encouragement of many high-class newspapers, amongst them the *Times, Saturday Review, British Medical Journal, Medical Press, Whitehall Review, Public Opinion,* and the *Provincial Medical Journal;* and we regret that we can give space neither for extracts, nor even a list of their names. The *Times'* notice commences as follows:—" A cordial recognition is due to the great public service rendered by the republication of the papers contributed to our contemporary by its editor." Such flattering praise gives us encouragement and increased energy to continue our work. We now present a second volume to our readers, and additional articles will appear from time to time in future numbers of HEALTH NEWS.

HEALTH NEWS Offices, Savoy House,
115, Strand, London. W.C.

CONTENTS OF VOL. II.

	PAGE
PREFACE	5

CHAPTER I.—" Pink Pills for Pale People " .. 13

CHAPTER II.—Warner's " Safe Cure " .. 22

CHAPTER III. — Quack Advertising; Clarke's Blood Mixture, and the Bogus Testimonial from the late Dr. Swaine Taylor, F.R.S. .. 33

CHAPTER IV.—Anonymous Abuse; Warner's Safe Cure, and Medical Staff; A Quack Libel Case; Morison's Pills; Baillie's, Dixon's, Fothergill's, and Lee's Pills 38

CHAPTER V.—Electric Belts; Mattei's "Electricities"; Nicholson's Patent Artificial Ear Drums 46

CHAPTER VI.—St. Jacob's Oil; Mother Seigel's Syrup; Mattei's " Electricities " 52

CHAPTER VII.—Our Correspondents and Critics; Silverton's Remedies for Deafness; Unqualified Practitioners; "A Merciful Medicine, more Precious than Rubies".. 59

CHAPTER VIII.—Patent Medicines and Pious Language; the "Reverend Specialist"; Congreve's Balsamic Elixir; Owbridge's Lung Tonic; Lane's Catarrh Cure; A Quack's Certificate 67

CHAPTER IX.—Stepney Green Diplomas .. 81

CHAPTER X.—Bone-setting; A Patent Medicine Song.. 88

CHAPTER XI.—Quack Bogus Newspapers; How the Poor are Swindled; Handyside's Consumption and Cancer Cure; Electric Snuff .. 93

CHAPTER XII.—Mattei's "Electricities in Court; A Curious Way of Exposing Quackery .. 106

CHAPTER XIII.—Patent Medicine Testimonials; St. Jacob's Oil; Clarke's Blood Mixture; the Man who gave himself a Testimonial; Eno's Fruit Salt 111

CHAPTER XIV.—"Handy" Still Shuffling .. 121

CHAPTER XV.—Quack Newspapers; The Sequah Bubble Burst; Translation of "EXPOSURES" for India 130

CHAPTER XVI.—Fenning's Fever Curer .. 134

Chapter I.

"Pink Pills for Pale People."

"Some to the fascination of a name
 Surrender judgment, hoodwinked."—Cowper.

"Words, words, mere words," as Shakespeare writes in *Troilus and Cressida*: catchpenny alliteration for an evident purpose. Of course the letter P may set some people thinking of paragons, pearls, and so forth; but we unhesitatingly assert—having regard to the cheap composition of the nostrum, the absurd claims of absolutely impossible remedial properties, and the way in which these pills are brought before the public—that we instinctively call to mind that P commences pence, profit, and pickings; "pence" representing the small cost of manufacture, while "profit" and "pickings" supply us with an explanation of all this puffing and pother about one of the commonest of drugs, as we shall presently show beyond all power of contradiction.

The "Pink Pills for Pale People" are also styled Dr. Williams' Pills, and in a pamphlet which accompanied our purchase—we hasten to explain that we buy such things for analysis, not for individual use, otherwise our series of articles might come to an abrupt conclusion—this mythical person is described as "an eminent graduate of McGill Medical College, Montreal, and Edinburgh University, Scotland."

Williams is by no means a rare name, so that the proprietors of the pills might have given his Christian one, in order to enable us to know more of this great man. At present we can only say that we have exhausted all means of research without being able to trace him in the graduation lists of either McGill College, or Edinburgh University. So far, we have the fullest reason to doubt the accuracy of the description of him by the pamphlet-writer for the Dr. Williams' Medicine Company; and our doubt ripens into positive scepticism when we read on the same page that the "Pink Pills for Pale People" are not a patent medicine, but are "a thoroughly scientific preparation, the result of years of careful study on the part of the eminent," &c. That they are a patent medicine, we very positively assert, and as a matter of fact they are sold with a patent medicine stamp attached to each box. What we do most unhesitatingly assert, too, is that they are not in any degree entitled to be described as a "thoroughly scientific preparation," or as "the only perfect remedy ever discovered," as alleged in the pamphlet. We find it also necessary, in the interests of truth, to expose the absolute falsity of the statement made a little further on, that "they supply in a condensed form, the substances actually needed to enrich the blood and restore the nerves." Bosh! utter bosh! as we shall presently demonstrate, when we have done with the pamphlet.

In order to invest this singular production with a degree of importance, it opens with a declaration made on oath before the Lord Mayor of London, introduced by these words:—" In order that the British public may know that every word in the remarkable narratives presented in the following pages is the truth, the whole truth, and nothing but the truth, we present the following sworn declaration made before the Right Honourable the Lord Mayor of London at the Mansion House, April 15th, 1893. In face of the sworn testimony, all doubt as to the marvellous curative properties of Dr. Williams' Pink Pills must be dispelled. Carefully read the evidence, and, if suffering from any disease arising from bad blood or shattered nerves-profit by the lesson it contains."

This is signed by the *business manager* of the Williams' Medicine Company, as well as the declaration, a very ordinary kind of document, alleging that, to the fullest of his knowledge and belief, the statements, names, and addresses printed in the pamphlet are true.

In accordance with the Statutory Declarations Act, 1835, anyone can go before a magistrate, or a commissioner for oaths, and make a similar declaration, the whole matter occupying a few seconds, and involving an outlay of a few sixpences. For instance, suppose a man writes a pamphlet to prove that the earth is flat and not round, there is nothing to prevent another man going before the Lord Mayor or any other magistrate, and making a solemn declaration to the effect that to the fullest of the knowledge and belief of No. 2, No. 1's statements are "absolutely and positively true." Consequently it will be seen that, to any rational individual, a thousand such declarations as that made by the Williams' Medicine Company's manager would be utterly useless for the purpose of proving the truthfulness of the reckless assertions as to the marvellous curative properties of the Pink Pills. On the other hand, any rational individual would regard a person so closely concerned, as the business manager must be, in the sale of this patent medicine, as the last individual who should come forward to assert the value of a nostrum in which he was pecuniarily interested.

As for Williams' Pink Pills, the Lord Mayor of London knows no more about them, actually, than the doorkeeper of the police-court at the Mansion House, where his Lordship sits for the transaction of public business (including the taking of declarations), or than the constable on duty outside. But it sounds grand to make a declaration before the Lord Mayor of London, and goes down with the unwary or the ignorant as something very important and impressive, and patent medicine men are not slow in availing themselves of handy methods of cheap showy advertisement to enable them to catch the coin of the multitude.

We have taken some trouble to look through this pamphlet, written in a clap-trap style throughout, for the alleged purpose of presenting to the British Public "the truth, the whole truth, and nothing but the truth" about the Pink Pills; of which pamphlet it is only right to say that the Lord Mayor of London never read a line, though his name is flaunted in the cover, and here and there in its pages, as carefully as if Sir Stuart Knill (who filled the position of Mayor in 1893,) had himself vouched for the accuracy of every word of the contents, highly coloured like its gaudy cover. The William's Medicine Company's business manager is evidently as destitute of humour as the pills are of the properties falsely attributed to them; otherwise, seeing the object of his declaration, he would have timed it a fortnight earlier, so as to fall on the 1st of April.

The perplexity attending our search for the name of Dr. Williams, inventor of the Pink Pills, in any graduation list of Edinburgh University, or of McGill College, Montreal, still pursues us when we come to the bewildering list of diseases which the business manager of William's Medicine Company parades before his readers. It goes almost without saying that the business manager claims for the patent medicine he is interested in, that it will cure fevers, consumption, paralysis, old age, and other alarming human ills; such assertions come naturally in a quack medicine pamphlet.

Beecham tells us, for instance in his trade circulars, that his pills will search out and remove all kinds of disease, "as sure as water quenches thirst."* But at that point his inventive power fails him; he can only describe in bad English and worse spelling, remarkable "*seeds or symtons* of *direful diseases*, like *hitching* of the breast or head." The Williams' Company business manager can give him points, for he enumerates in the list of diseases which Pink Pills search out and infallibly cure, some which have never yet found a place in medical books, namely, "lack of ambition" and "shallow complexion." What a chance offers for some new Barnum to exhibit a sufferer from this last-named disease! He must, however, in the words of Mrs. Glasse's famous cookery book, "first catch his hare," and if he searches only amongst patent medicine men he will never succeed in obtaining a specimen, for they are too deep to suffer from "shallow complexions."

The pamphlet tells us everything about the Pink Pills—everything that is to say, calculated to promote their sale—but, while concealing the full name and address of their alleged inventor, it also conceals the composition of the pills; or rather, which is worse, it gives very many completely untrue statements. As to their being the "only scientific and rational and only perfect remedy, ever discovered," we will not insult the common sense of our readers by taking further notice of such obvious falsehoods made more glaring by appearing in a pamphlet which the Lord Mayor of London is asserted to have allowed to be placed

---

* See Exposures of Quackery, Vol. I., page 96.

on record as being "the truth, the whole truth, and nothing but the truth." Where could the business manager have learned the physiological fiction contained in the assertion that the Pink Pills "supply, in a condensed form, the substances actually needed to enrich the blood and restore the nerves?" Did Williams tell him this nonsense when selling his wonderful secret to the company? If we may gauge Williams' knowledge by this sample, there is a far greater miracle than the pills to be accounted for, and that is, how Williams ever qualified himself to become—though we don't, for a moment, believe that he ever did become—"an eminent graduate" of that old-established seat of learning, Edinburgh University. Such assertions "won't wash," any more than the pills themselves, which soon loose their pretty pink colour when placed in water.

We defy anyone living to produce a treatise on physiology which mentions the component parts of these pills amongst "substances actually needed to enrich the blood and restore the nerves." Our readers may begin to ask what these component matters are. Well, as neither the Lord Mayor of London, nor the door-keeper of the Mansion House police-court, nor even the outside constable are here to give us information, if they could, we must fall back upon a more reliable source, and invoke analytical assistance.

Mr. George Selkirk Jones, an analyst of many years' standing, and author of the *Chemical Vade-mecum* has examined them for our journal. We append his report:—

"PINK PILLS FOR PALE PEOPLE."

"I have now made a careful analysis of these pills, and I find their composition to be as follows :—

"Extract of Barbadoes *Aloes*, enclosed in a thin coating of *Sugar*, coloured pink with *Carmine*.

"Seeing that these pills are said to have been successfully used in America for the cure of a 'given-up' case of paralysis, and also of rheumatism, fevers, &c., I have carefully examined them for other drugs, but have discovered none other than that mentioned, viz., Aloes. If asked for my opinion (as a medical practitioner) whether these pills are capable of doing what is stated of them, *upon oath*, I should answer emphatically, 'No, certainly not.'"

Our old acquaintance, again, Aloes, the universal sheet-anchor of patent medicine makers; the commonest kind of aloes, too, namely, the Barbadoes species, the best suited for horse balls and cattle physic, according to veterinary authorities. This vulgar form of a very common drug is the sole medicinal agent upon which the mysterious Williams bases his claim to rank as a leading scientist and a great discoverer. Why, even his patent medicine brethren would dispute his pretensions. Mother Seigel, the Indian Sequah, Holloway, Beecham—one and all—assert that they discovered aloes, though this drug, was known centuries before patent medicine men had any more existence than venomous reptiles in Iceland. Good old times!

Perhaps, since aloes will not cure consumption, paralysis, fevers, &c., Williams, "the eminent," may rely on the other ingredients. He is welcome to any small comfort he may derive from them; the sugar to sweeten his chagrin at the exposure of his pretensions, or the carmine with which they are coloured to hide his blushes. We as much doubt, however, a patent medicine man's ability to blush as we do the alleged power of this nostrum to perform miracles; and further, there is very strong internal evidence that the "eminent" Williams is as thorough a creation of fiction as the "eminent" Ally Sloper, of comic newspaper fame, or Mrs. Gamp's never visible friend, Sairey Harris.

We chanced to see last April, (this is a favourite month in the Williams' Calendar, apparently), the following remarks, carefully worked into a column of a religious paper, as if it were a piece of news:—

"MEDICAL FRAUDS.—CAUTION. No imposition can be meaner than foisting upon a sick man a worthless substitute for the medicines that will restore him to health."

All of our readers will readily coincide with the opinion here expressed, but they will be much astonished when they learn that this quotation serves as a mere "draw" to induce people to peruse a mendacious advertisement of "Pink Pills for Pale People." We need not look further than these pills for an object-lesson on the meanness of foisting worthless quack medicines on unfortunate sufferers. We admit, that for once, the author of the Williams' puffing literature has so far forgotten his vocation as to speak the truth.

Chapter II

Warner's Safe Cure.

"*Villainous saltpetre.*"—Shakespeare.

The *Saturday Review*, in a very long and eulogistic article, comments upon our exposures of popular patent remedies.

"The method of exposure employed," says our excellent contemporary, "has been simple yet drastic. The nostrum has been submitted to analytical examination, and in each case has been discovered to be a preparation of well-known ingredients, *well-known* not to possess the properties claimed by the vendors for their secret compositions. 'Clarke's Famous Blood Mixture or Purifier' for instance, consists, says Mr. Stokes, the public analyst, of iodide of potassium, chloric ether, potassium hydrate, and coloured water. To claim for this combination that it is a never-failing and permanent cure for scrofula is—to put it with a decent restraint not noticeable in Mr. Clarke's advertisements—a little extravagant. The assertion, therefore, made by the proprietors of the Blood Purifier, apparently without fear of contradiction, that that medicine is the only cure for consumption, diabetes, dropsy, deafness, and paralysis, is an over-bold one. 'Mother

Seigel's Syrup,' 'Sequah's Prairie Flower,' and 'Holloway's Pills' can jointly cure everything and can severally make a good job of most things (at any rate so their manufacturers unhesitatingly suggest). But the analyst has reported on them, and for the future in unimaginative minds they can only be credited with the virtues of their one active ingredient, Aloes. The triumph of the editor of HEALTH NEWS has been complete."

The *Saturday Review* goes on to descant upon the rise and fall of quack remedies, and the difficulties in the way of attempting "to make laws for the effective protection of the pockets of the gullible." "Quacks have been, and quacks will be, and there will always be a public ready to heed them, and happy to pay them," observes the *Saturday Review*. "Admitted," we reply; but our argument is that the British Government, for the sake of obtaining an annual addition of some £200,000 or so to the national income, by the issue of patent medicine stamps, tolerates and even endorses quackery; for many, and especially ignorant, people— not always of the lower classes—are foolishly deluded by the Government Stamp and the word "Patent" into a belief that they convey a sort of guarantee of quality. As to laws for protecting the gullible, the unwary, and the ignorant, such laws exist in connection with every other mode of imposture. Nor would any elaborate legislation be needed. If an individual buys a pound of coffee, to which chicory has been added, he has his legal remedy under the Adulteration Act, unless the dealer has previously placed upon the packet a legibly printed label showing that the contents are not pure coffee, but coffee and chicory mixed. In the same way, instead of shrouding a quack medicine in mystery—*Omne ignotum pro magnifico* is the theory of many people—and lending it a fictitious value by affixing an official stamp, let the Government pass a short Act of Parliament, similar to the legislation which prevails in various Continental countries, requiring the contents of every bottle, box, or packet of quack medicine to bear a label stating its real composition. People are getting more educated than they used to be, and education is the deadliest foe quackery can have. Would any man or woman of sound mind and even very moderate education, if he or she learned from the printed label on a patent medicine, that it consisted of water, aloes, and carbonate of soda, with a few drops of the tinctures of capsicum and myrrh,* be likely to give credence for one moment to the preposterous assertion of its vendors that "thus compounded" (we copy this statement *verbatim* from the prospectus accompanying a bottle of Sequah's stuff, sold at the rate of a shilling an ounce, (seven-eighths being water, and the other ingredients the commonest and cheapest of drugs) "PRAIRIE FLOWER" is undoubtedly far and away the best remedy ever introduced for all sorts of complaints and other CHRONIC DISEASES." The capitals in this singularly worded extract are, we need scarcely say,

---

* See analysis of Sequah's Prairie Flower Mixture, in EXPOSURES OF QUACKERY, Vol. I., Page 40.

not our own, but those of the compiler of the prospectus who evidently acted on the vulgar rule that when anyone tells a lie he should tell a big one, and stick to it.

There was once an American quack pill, whose discoverer and inventor—quacks are remarkable people, for they invariably discover and invent, according to their own version, things which have been known ever since the world began—claimed for it the meritorious qualities that it didn't "go fooling around, but settled down steadily to business"; and we fear that our readers will accuse us of the one, and charge us with not doing the other, unless we forthwith say something about Warner's "Safe Cure." "Safe Cure," indeed! That is what its inventor and discoverer would probably, in his Trans-Atlantic vernacular, call "a tall order," if the nomenclature emanated from anybody else. Still, it is not absolutely original; years ago there flourished a music-hall celebrity, one Mr. Stead—we hasten to explain; not the gentleman of that name who poses as the champion of Mattei and his watery electricities, white, red, and green!—who jumped himself into fame as the "Perfect Cure."

We have before us an analysis of the "Safe Cure for Bright's Disease, etc.," of which we will give particulars. During a recent illness, as at other times, Shakespeare has been a frequent companion, and it was when reading one of Shakespeare's works that Warner's "Safe Cure" came into our mind. An extraordinary concurrence of ideas, some will remark, and difficult to account for ; yet they should remember that it has been paradoxically asserted that the improbable often becomes the possible. Macbeth thought himself on the safe side when he more than hinted at the physical impossibility of the removal of Birnam Wood to Dunsinane, but he had to own up to his mistake very soon afterwards. Which play of Shakespeare was it that suggested the " Safe Cure " ? Our publishers do not find it necessary to stimulate the circulation of HEALTH NEWS by giving away pounds of tea or other bonuses to subscribers, or offering conundrums for competition, so that there can be no excuse for delaying the answer —" King Henry IV.," wherein Hotspur makes mention of his prisoner's decided objection to " villainous saltpetre." Of course, seeing that the prisoner had just run a narrow risk of losing his life in the battle, the epithet he made use of was excusable; otherwise it might seem rather too strong a term to apply to what is, and has been for centuries, quite a common article of commerce. How would Colonel North, the uncrowned Nitrate King, like to hear anyone speak of nitre in such uncomplimentary language ? But we did not happen to think of him. We thought of Warner, and of what a lot of saltpetre there must be in his " Safe Cure."

We were right, too, for an analysis specially made for us of this much puffed medicine, by Mr. A. W. Stokes, F.C.S., F.I.C., public analyst, revealed the fact that three hundred and fifteen grains of saltpetre were contained in a sixteen-ounce bottle. Sixteen ounces! Rather a stiff quantity of physic, this. Warner and Co. think so too, evidently, for they make the boast, "Our bottle is the largest 4s. 6d. bottle in the market." We will not attempt to disprove this assertion, but we should have preferred its being half the size and half the price, as we bought it, not for home consumption, but with the view of sending it to our analyst. If a "largest bottle" craze should seize upon the nostrum-loving public, we may expect eventually to see further developments in this direction on the part of other quack medicine vendors, such as notifications announcing "on tap," "in the customer's own jug," and "small casks for family use."

The dose recommended by Warner and Co., is a tablespoonful, *i.e.*, half an ounce, six or eight times a day. Eight doses would equal four ounces; four ounces, multiplied by four, equal sixteen ounces, and hey, presto! in four days the largest bottle in the market has been emptied, and gone with its contents are four shillings and sixpence, good and lawful coin of the realm.

Now, what has the purchaser had for his money, besides 315 grains of nitre, *alias* saltpetre, *alias* nitrate of potash, the value of which anyone, curious on this point, can ascertain at the nearest drysalter's? Why, Mr. Stokes tells us that, in addition to water and the aforesaid saltpetre, he extracted from the sixteen ounces of fluid one-and-a-quarter ounce of glycerine, half-an-ounce of burnt sugar or treacle, two ounces of rectified spirits, a few drops of oil of winter-green, and a vegetable extract bearing resemblance to extract of liver-wort. He could not find any alkaloid, or any of the usual drugs employed in the treatment of Bright's disease. "My belief," he adds, "is that the nitre is the only active ingredient present."

We fully endorse Mr. Stokes' opinion, and we also believe that any person who would attempt to treat such a fell disease as that which Warner and Co. profess to eradicate with either saltpetre or glycerine, or burnt sugar or treacle, or rectified spirits, or oil of winter-green, or extract of liver-wort (the two latter in such small quantities that we should have to seek the aid of fractions to enable us to calculate the amount taken for a single dose) would not have such great difficulty as Dogberry experienced, in getting himself written down an ass, in Shakespeare's play of "Much Ado about Nothing."

If winter-green and liver-wort do not show up in any large proportions, the same cannot be said of spirits, which constitute one-eighth part, or $12\frac{1}{2}$ per cent. of the whole quantity; yet there is no disease in which more care should be exercised as to the use of alcohol than in Bright's disease.

It goes almost without saying that Warner and Co. issue, broadcast, circulars dilating upon the paramount necessity of everyone who feels out of condition at once proceeding to drug himself with their preparation.

"The doctors cannot cure you—this they admit," Warner and Co. dogmatically affirm in one of these pamphlets. A bad look-out for every man or woman who is, or fancies that he is, or she is, not in health. Yet, there is a silvery lining to every cloud—or rather, in this instance, a particularly brassy one—for this consolation to the afflicted speedily follows:—"Treat yourself with Warner's Safe Cure, and live. All that the medical faculty can do is to make dying people comfortable. Thousands of people die every year from supposed apoplexy, convulsions, heart disease, paralysis, gangrenous erysipelas, and other quick-ending disorders, when in reality they are the victims of chronic Bright's disease. Their physicians cannot cure it; and they, therefore, to cover their inability, attribute death to other causes." And so on, till we are lost in amazement at the mean, despicable falsehoods that some people will be guilty of, to make money; for the object of these pamphlets, dropped into family letter-boxes, and given away indiscriminately, is to promote the sale of the "Safe Cure."

One of these pamphlets, left at our private house some time back by a man who was distributing them throughout the district, contains an earnest invitation to those into whose hands it may fall, to forward by parcels post a six-ounce sample bottle of their urine to H. H. Warner and Co., Limited (Medical Department), for examination; "the charge made being 2s. 6d., barely sufficient to cover the cost of chemicals."

What is a humble half-crown compared with "long practical experience, involving the examination of *many thousand* samples annually"? But how can Warner and Co. get through this amount of chemical and microscopical research, and to whom do they entrust such responsible duties? Mr. Warner's personal services cannot always be relied on.

Even if his health, however much fortified by nitre, syrup, etc., did not break down under the continuous strain imposed on him, he must at times be absent. It is not long since we read in an American paper that he was then at Rochester, State of New York, superintending his "mammoth yeast" business, and making arrangements for bringing out Warner's "Safe Baking Powder"; which he assured an interviewing reporter, who straightway blazoned the circumstance in a "mammoth" advertising article, would be as far in advance of the powders now upon the market" (the "largest bottle" is already there) "as the 'Safe Remedies' are above the vile imitations and substitutes which are offered in their stead."

Really, after what we have learned from Mr. Stokes' analysis, we would just as soon have vile imitations and substitutes as "villainous saltpetre." Do the English directors of H. H. Warner and Co., Limited, conduct many thousands of urinary analyses annually? It is hardly likely, we should imagine, that they would undertake scientific experiments on such a gigantic scale. Certainly not, and, on looking through the pamphlet again, we come to the conclusion that "the physicians employed in our Medical Department" must be the persons who, during Mr. Warner's unavoidable absence, owing to the demands of yeast and baking powder on his attention, carry on the laboratory work. This does seem cool, not to say cruel, conduct. In one paragraph the sufferer is told in the most positive language that doctors cannot, and admit that they cannot, cure him, and at the best they can only enable him to die comfortably; in another, that "*our* physicians will gladly give the benefit of their knowledge, free of cost." If we believed anything Warner and Co. said, we should estimate the benefit (?) of their physicians' knowledge as appraised in the preceding sentence at its exact value, viz.—*Nil.*

But their "physicians" are not the guileless philanthropists that Warner and Co. would have us imagine. A letter was shown to us not long since, written to a young man by "our physicians," or one of them—in their impatience to benefit humanity, Warner and Co. have omitted to inform the public as to the strength of their medical staff, or even to give the names of their doctors—in which the following paragraph occurs:— "Charge for the month's treatment is £4 4s. If a second month's treatment is necessary, charge is £3 3s." With this letter was enclosed a printed consultation form, containing some of the nastiest suggestions that could be put into a youth's head. Presuming that the fee was considerately dropped every successive month, in the same ratio, our young friend calculated that the fifth month's treatment would have been literally "free of cost." But it will never reach that stage, and even the Parcels Post six-ounce package and the half-crown for bare expenses, will never pass the portals of H. H. Warner and Co., Limited. We cheerfully gave the benefit of our knowledge and advice, free of cost, as they would say, in the single word indelibly engraved on our memory by *Punch* in his advice to young people on another subject,—DON'T; and we had the satisfaction of knowing that the quacks had one dupe the less, through our timely intervention.

How many more the publication of our article has saved from loss of money, time, and health, we cannot of course, say. At the date when it first appeared— some few years ago—Warner and Co., Limited, were paying a dividend of 17½ per cent. This grand harvest from the ignorant and the unwary has since dropped to *Nil*, and the shares in the Company are now as unsaleable as the "Safe Cure" is becoming. Our readers will doubtless draw their own conclusions from these significant facts.

Chapter III.

Quack Advertising; Clarke's Blood Mixture, and the Bogus Testimonial from the late Dr. Swaine Taylor, F.R.S.

We have frequently had occasion in this series of articles to refer to the persistent and perpetual puffery of quack nostrums. Like the poor, mentioned in Scripture, the "pillionaires," as Mr. George R. Sims has designated the people who make hundreds of thousands, even millions, of pounds out of advertising aloes, or some equally cheap, common drug, as capable of curing every human ill, are for ever obtruding themselves upon us. Whether walking, riding, eating, drinking,—whatever we may be doing, in fact—we get ocular evidence of the ceaseless activity of the quacks. Why, one is almost led to suppose that they never go to bed like ordinary folk, for they are in evidence from our rising up to our retiring to rest; and, if perchance we should need to strike a match in the middle of the night, the odds are ten to one against our not finding the box decorated with the statement that Gullaway's Ointment or Fleece'em's Pills will confer long life and the best of health upon anyone idiotic enough to be taken in by such reckless assertions. Mr. George R.

Sims, in a recent *Referee*, strongly denounced, in his customary humorous style, the extent to which quack advertising had gone; the latest development of it being the erection of huge boards alongside the railway lines, painted in bilious-looking or bright crimson colour. "Somebody's Pills hold the field," would seem to be the highest literary effort of the men who thus insist on adding a new misery to railway travelling. Of course, we shall be told that such a proceeding is strictly legal, that anyone—provided only that he arranges with the owner or occupier of the land,—can erect the most hideous hoarding he chooses, and cover it with the most objectionable and untruthful statements. "Did you see all those quack advertising boards standing alongside of the line as you came down?" inquired a friend whom we were visiting in Surrey; as if, indeed, any one not absolutely blind or, previously to starting, under the influence of chloroform, could fail to see them! "We saw a number of quack advertising boards *lying* alongside of the line," was our reply, delivered in such an emphatic manner that our friend looked for the moment as if he thought the wrong man had accepted the invitation to spend the day with him. Mr. Sims tells an amusing story of a romantic young lady, who travelling in the same carriage with him, closed her eyes, and said to her mamma, "Mother, tell me when the scenery isn't all pills, and I'll look at it again." Then Mr. Sims expresses in verses his disgust at this general debasement of the English landscape for advertising purposes.

Somebody's Pills.

The sun o'er the valley is streaming,
    The lambkins are frisking with glee;
In the bright light the rivulet gleaming
    Meanders away to the sea.
The meadows with daisies are dotted
    And crowned as with gold are the hills;
But the whole of the landscape is spotted
    With advice to try "Somebody's Pills."

The woodland is brave in its glory,
    The fulness and freshness of spring;
Round the castle afar, old and hoary,
    The ivy leaves clamber and cling;
There is rest for the eye as it gazes,
    There is joy for the heart as it thrills
But the beauty is all sent to blazes
    By the big boards of "Somebody's Pills."

If the traveller, wearied and annoyed at the way in which these quack advertisements are paraded at every point of view, turns to his newspaper for some relief, even then the ubiquitous quack announcement meets his eye, and vexes his spirit. Not unfrequently, too, he comes across therein some strange heading, such as "Ten minutes more, and he would have been beyond help," "Saved by a String of Sausages," "A Night in Newgate," and similarly startling titles; he reads a few lines, becomes interested, then drops the paper with an inward groan as it flashes across his mind that the preliminary paragraphs only lead up to a barefaced puff of Grandmother Seagull's Treacle, or some other nostrum, which like *Peter Pindar's* Razors, are "made to sell." His feelings, as the truth dawns upon him, are well portrayed in the following lines, which appeared some time since in a London weekly paper.

ON PATENT MEDICINE STORIES.

It was only a newspaper story,
   And yet, as I read it o'er,
My eyes grew moist and heavy
   As they had not in years before.

It was not the art of the writer
   That on my heart-strings swept,
But the story simple and tender,
   Went to my heart as I wept.

But when I arrived at the "finis,"
   It caused my heart to ache;
And I spoke strong words, for that tender tale
   Was a patent medicine "fake."

Talking of quack advertisements, we are reminded of the sudden disappearance from the London daily and weekly papers of one about which we have had a good deal to say, in this series of articles in HEALTH NEWS. We refer to the testimonial widely advertised by the proprietors of Clarke's Blood Mixture, and alleged to emanate from the late Dr. Swaine Taylor, F.R.S. We have published irrefutable evidence that no such testimonial could ever have come from that able analyst and distinguished scientist, and we have stated our firm belief that the alleged testimonial was either a myth, or an impudent forgery. But the proprietors of Clarke's Blood Mixture have been suspiciously silent; they have withdrawn the bogus testimonial from their advertisements; but they have not ventured on the least explanation, although each of the various numbers of the journal containing our comments on the matter has been sent to them by registered letter. They have "climbed down," and scuttled off in a manner on a par with their flagrant abuse of the honoured name of a dead man.

But they must not imagine that they can escape scot-free in this way. We reassert, with the emphasis of full conviction, heightened by a sense of duty to Dr. Swaine Taylor's memory, that the alleged testimonial advertised by the proprietors of Clarke's Blood Mixture is either a myth, or a forgery; and we are prepared to make a wager of any moderate amount—not belonging to the class of "pillionaires"—that our assertion is correct. It would be repugnant to our feelings to receive Blood (Mixture) money; so that we must make one stipulation, namely, that the winner shall divide the proceeds of the wager between deserving medical and masonic charities.

They may deem it prudent to maintain silence, but we do not intend to let the matter drop. It is neither honourable nor honest on their part to cast a slur on a dead man's reputation, and then to refuse a word of explanation.

Chapter IV.

Anonymous Abuse; Warner's "Safe Cure and Medical Staff; A Quack Libel Case; Morison's Pills; Baillie's, Dixon's, Fothergill's, and Lee's Pills.

"*Who is this, that deafens our ears with abundance of superfluous breath?*"—Shakespeare.

Since the commencement of this series of articles, we have received so many abusive and threatening communications, apparently from persons connected with the patent medicine trade—in their anger or their modesty (?) they have omitted to give their real names and addresses—that we are getting quite used to letters of this sort. Indeed, we are becoming hardened, and ready to accept them as proofs, notwithstanding their invariably uncomplimentary language, that these articles are too truthful to be palatable to some people, at least. Well, we started on the old-fashioned plan of calling a spade a spade; and when we consider the extravagant boasts which our anonymous assailants make concerning the often purely imaginary properties claimed for their wares, we cannot help thinking it would be better if they would go straighter to the mark, and not perpetually walk round the truth.

There is another curious coincidence about these communications, besides the abuse, energetic enough to make the fortune of a professor of (bad) languages, namely, that they all propound questions which they do not themselves seem able to reply to. Even if, as the writers apparently imagine, abuse is argument, questioning, without supplying the answer when needful, is a low form of logic. One correspondent inquires: "Why should not Warner and Co. have a large medical staff composed of experienced physicians?" We never said they should not; in fact, a pamphlet dropped into the letter-box at our private house distinctly states that they are fully equipped in this respect. If our correspondent will look again at our article on "Warner's Safe Cure," he will see that we not only refer to this statement, but that we exhibit a natural curiosity as to the names and professional position of the physicians of "long experience and extensive practice" who are always, like Mr. Micawber, waiting for somebody or something to turn up at Warner and Co.'s, where "consultations by letter or in person are invited, and medical advice is cheerfully given without charge." This may, or may not, be the case.

The pamphlet assures us of it, as well as of "strictest confidence;" but, in illustration of the old adage that there is an exception to every rule, the only experience we have had of Warner and Co.'s medical staff and their cheerfully gratuitous advice is derived from a letter which was shown to us by a young man who was induced to write to Warner and Co.'s physicians. The reply he received urged him to fill up a special printed form, comprising no less than forty questions, many of them absolutely unfit for publication. The fee demanded for the first month's treatment was £4 4s.; for the second month, £3 3s. This does not carry out the liberal offer made in the pamphlet, nor does it enlighten us much as to the composition or number of the medical staff.

The letter is signed by an individual who styles himself M.D. Pa., U.S.A., a degree unrecognised in this country. The following is an extract from an official answer to our inquiry on this subject addressed to the Registrar of the General Council of Education and Registration of the United Kingdom :—"The qualification of M.D. Pa., U.S.A., does not entitle to registration in the *Medical Register.*" The reason is not a difficult one to solve.

What a ridiculously small medical staff it would be were it composed only of one M.D. Pa., U.S.A.; but, of course, Messrs. Warner & Co., must know what they write about, and they mention "physicians." Plural, be it observed; a plural number might constitute a staff, though one would not. If Warner & Co. are really too preoccupied with baking powders (what wonderful "puff" paste they ought to turn out!) to give the names of their physicians just at present, perhaps they will obligingly remedy this omission in the next million of their pamphlet.

We have carefully looked through the circular, and find no information on this important point. We came across a testimonial purporting to emanate from a medical practitioner. It is as follows:—"91, Hoxton Street, London, N., August 12, 1889.—The greatest recommendation that I can give to Warner's 'Safe' remedies is their vast increasing sale, showing their undoubted worth.—M. E. WILLIAMS, M.D., &c." We thought it strangely worded, though, giving such evidence of the writer's acquaintance with the details of Warner & Co's business that we could not help arriving at the conclusion that the writer might have derived his knowledge of the "vastly increasing sale" from holding a position on Warner's medical staff.

We are again placed in a dilemma, for a close examination of the *Medical Register* and of the *Medical Directory*, containing the names of more than 30,000 qualified practitioners in the United Kingdom and abroad, failed to discover "M. E. Williams, M.D." amongst that large number, any more than the M.D. Pa., U.S.A., who proffers "*gratuitous*" professional advice at the rate of four guineas per month! Surely, Warner & Co., in their overweening faith in saltpetre, have not adopted such a curious method of demonstrating their disregard for a profession, which their "Safe" Remedies are destined, in their own opinion, to clear off the surface of this planet, as making it a *sine quâ non* that no qualified doctor be deemed capable of acting upon their medical staff. If such is the case, they ought, in fairness, to make the fact public. Will M. E. Williams enable us to rectify the omission on the part of the official *Medical Register?* Despite his "greatest recommendation" of Warner's "Safe" remedies, he assuredly thinks there is a corner left yet for medical commercial enterprise; for, in the chemist's window at 91, Hoxton Street, we saw displayed sundry bottles of what Williams modestly styles his "Lung Restorer," and a "Blood Purifer, specially prepared by M. E. Williams, chemist, late army surgeon."

On the principle that "one good turn deserves another," or, as the Scotch saying runs, "Ca' me, and I'll ca' thee," Warner and Co. should give M. E. Williams, &c., a testimonial asserting that he is just as great (and no more) in purifying the blood and restoring lungs as they are in safely and certainly curing with saltpetre diseases beyond the limited resources of the whole medical profession.

Some curious evidence came out not long since in a trial at the Cheshire Assizes, being an action brought by Dr. Alfred Ellis Vaughan, a medical practitioner at Crewe, against Samuel Johnson, of Wrinehill, near Crewe, "a quack of the purest water," as the Liverpool and Cheshire papers describe him.

This Johnson though fit to publish statements of an undoubtedly and grossly defamatory character concerning Dr. Vaughan's treatment of a patient, who in some way, fell into Johnson's hands subsequently. The pamphlet contained most extraordinary puffs in praise of what Johnson called his "Chinese Pills;' and Johnson coolly announced that he could cure the worst fever in three days certain, the worst inflammation in six hours, diphtheria in six hours, brain fever or inflammation of the brain in twelve hours, the worst quinsy in two days, and stoppage or twisting of the bowels in six hours certain. There was, in short, no disease that could withstand the Chinese remedies. Marvellous, *if* true (what virtue lies in that little word "if!"). Indeed, there could be only one thing more marvellous, and that is how any man or woman in the county, outside the Chester and Macclesfield Lunatic Asylums, could be induced to believe such assertions. We cannot speak as to the composition of Johnson's Chinese pills; but we can state one fact with satisfaction, namely, that the jury gave a verdict in favour of the plaintiff, £250 damages, with costs. Whether Dr. Vaughan was paid the damages and costs after the trial we cannot say, but we fear there is as much room for doubts on this point, as on the Chinese pills.

As to these Chinese pills, we should not be surprised if they contained aloes, for two reasons:—(1) Quacks have a remarkable predilection for this cheap purgative, as will be seen upon a perusal of our previous articles; (2) they are also invariably wrong in localising the countries whence it is obtained.

Morison's Pills, specially prepared, of course, at the institution in the Euston Road, dubbed the College of Health, contain a considerable amount of aloes.

No. 1, Morison's, is composed (according to the analysis of Mr. Henry Beasley) of aloes and cream of tartar in equal proportions; No. 2 pill consists of two parts of gamboge, three parts of aloes, one part of colocynth, and four parts of cream of tartar, worked up into pills with the aid of syrup. Nothing wonderful or novel about these pills, at any rate, except that one greatly wonders what there is about them to render it necessary for the proprietors of the nostrum to style their emporium the College of Health, or themselves Hygeists.

Apropos of aloes, this forms a chief component of most quack medicines. Of these we have already described a number in this series of articles, including Beecham's Pills, Holloway's Pills, Mother Seigel's Syrup, Sequah's "Prairie Flower," and "Pink Pills for Pale People." Here is a batch of four analyses of patent pills in support of this statement :—

1. Baillie's Pills.—Extract of aloes, 1½ drachm; compound extract of colocynth, 1½ drachm; Castile soap, ½ drachm; oil of cloves, 15 drops; make up into three dozen pills.

2. Dixon's Antibilious Pills.—Equal parts of aloes, scammony, and rhubarb, with the addition of a small quantity of tartar emetic; Castile soap, to make up the mass.

3. Fothergill's Pills.—Aloes, antimony, scammony, and extract of colocynth.

4. Lee's Antibilious Pills.—Aloes, 12 parts; scammony, 6; gamboge, 4; jalap, 3; calomel, 5; soap, 1; syrup of buckthorn, 1: and gum mucilage, 7 parts; mixed together, and divided into five-grain pills.

Of a truth, quacks are not over-burdened with inventive genius, always excepting as regards their advertisements and testimonials.

Chapter V.

Electric Belts; Nicholson's Patented Artificial Ear Drums; Mattei's Electricities.

"*Gullible, however, by fit apparatus all Publics are; and gulled with most surprising profit.*"—Carlyle.

When the Sage of Chelsea penned these lines in *Sartor Resartus*, he must have enjoyed a cynical chuckle over the Publics who allow themselves to be taken in by the swindling "arts of Puffery and of Quackery," the "grand over-topping Hypocrisy," as he writes further on. Little wonder, indeed, is it that he reckoned up these blind believers in quack advertisements and nostrums as "mostly fools!" By "fit apparatus" he meant, of course, the means devised for trapping such people as the large class of the public who put faith in the patent *alias* quack remedies.

For our present purpose, however, we will take the word "apparatus" in the narrowest sense. One of our correspondents, speaking of electric belts, describes the case of an unfortunate man, a labourer, dying from cardiac dropsy, and so poor that he could barely get food enough of the humblest kind, who was persuaded by his credulous neighbours to lay out his little all in the purchase of the much-vaunted and extensively advertised electric belt of an Oxford Street firm. Of course, the man derived no benefit from his purchase, and soon afterwards died. Our friend got permission from the relatives to examine this *precious* apparatus, after the decease of the poor dupe. What did it turn out to be? Why, half a dozen discs of tin, as big as a florin, neatly sewn into a flannel belt, not worth as many farthings as the poor fellow had been robbed of shillings.

Yet these fradulent apparatus are advertised in the leading papers, some of them journals which would decline to insert announcements of most of the ordinary quack remedies in vogue. One of these papers, for instance, had not long since, in its advertising columns, a long rigmarole about electric treatment, commencing with a "magneto-galvanic sleep promoter," at two-and-a-half guineas, and winding up with "the electro-spiral hood to keep life in patients dying from exhaustion until the treatment has time to take effect. Price £500!" Great Scott! Which is most to be wondered at—the mendacity of the advertisers, the folly of the purchasers, or the weakness of the publisher of a high-class paper in allowing its columns to be made the medium of such a palpable imposture? And this sort of thing goes on day by day, week by week, and month by month, in hundreds of papers bearing a high character for respectability and veracity—religious papers being the greatest offenders, a circumstance from which an unbiassed observer might deduce the conclusion that their readers are not of the most intelligent classes — until one blushes for journalism. Indeed, the electro-magnetic charlatans find ink and paper such a profitable investment that they run papers of their own, for the sole purpose of puffing their own wares, while at the same time they make desperate onslaughts upon each other. Thus, *Modern Medicine*, a monthly periodical published in London, endeavours to convey to its readers that Matteism and miracle working are convertible terms, and wild assertions are made that the Mattei electricities* will promptly cure every known malady.

But let us shift the venue from London to Geneva, and we find there a so-called Electro-Homœopathic Institute, whence issues a monthly magazine printed in English, devoted to electro-homœopathy and to damaging attacks on Mattei, whose former agent, a Mr. Sauter, owns the paper and an opposition shop. Some of these are not wanting in humour. For instance, a letter is given, written by Mattei, to a newly appointed agent, showing that Mattei more than makes up for his want of scientific knowledge by his superabundant bombast. The translation is as follows:—"Assuredly you will not lack gold in millions. You will attract it as a magnet attracts

---

* The analysis of these by Mr. Stokes, F.C.S. sums them up as— *Water*. " nothing more."—See EXPOSURES OF QUACKERY, Vol. I.. Page 16.

iron. You have the magnet in your intelligence and your activity. Farewell, my dear sir; work with your ability, and I guarantee a good result." How is that for high? As a specimen of lofty rhodomontade, we know nothing to equal it. Hannibal crossing the Alps, Drake setting out to encounter the Spanish Armada, Napoleon on his march to Moscow, could scarcely have used more inspiring words. Yet, though they have a ring about them, it is a false ring, brassy in sound, inspired only by a desire for filthy lucre. But, as Burns said, "The best-laid plans of mice and men aft gang aglee." When the agent had spent much money in advertising—that goes without saying, as it is a quack medicine we are writing about—and had given much time and trouble to the business, Mattei picked a quarrel with him, refused to pay the agreed commission, and was, says his agent, Mr. Sauter, guilty of much meaness and injustice. However, this kind of treatment of his agents, judging by the numerous instances quoted by Mr. Sauter, is as much part of the Mattei method as the globules are. The only pity is that these rivals do not imitate the famous Kilkenny cats, and annihilate each other, instead of perseveringly obtaining the sinews of war from their deluded followers.

Passing from electric to aural apparatus, we may refer to a much-advertised instrument guaranteed for the complete eradication of deafness. A few years ago, an acquaintance, a London solicitor, consulted us for deafness. In the course of the consultation it came out that, attracted by an advertisement, he despite his profession which ought, like freemasonry, to have taught him to be cautious, was "fool enough" (we quote his own words) to spend—we forget how many—guineas upon the purchase of "Nicholson's Patented Artificial Ear Drums," *gold* of course! Finding his hearing seriously impaired by their use he had flung them aside. They may be described as follows :—Two straight pieces of gilt brass wire, three-quarters of an inch long, one-sixteenth thick, with a roughly made knob and a disc of very thin india-rubber, half-an-inch in diameter, at each end; a little loop of silk cord to hold each by. The purchaser is directed to wet one of these instruments with a lotion (charged extra) containing glycerine, ether, and morphia, and to thrust it into the ear, "until the natural drum is reached, and the end with the large disc well set into the outer ear."

It makes one shudder to look at it, and, still more, think of the mischief, aural and cerebral, it may do, and is certain to do, if left in the ear. The injury caused to thousands of people who were inveigled by advertisements and testimonials into investing ten times as many thousands of pounds in such worthless, dangerous apparatus, must have been incalcuable. Yet Nicholson was allowed by the police and by the Medical Council to go on for a considerable space of time (several years) advertising in metropolitan and provincial newspapers for dupes, from an address in one of the principal squares of central London, more often than not styling himself "Dr." Nicholson. Eventually he formed a limited liability company, drew most of the cash, and then vanished.

There are laws for the repression of theft, fraud, and imposture. Why should not the authorities do their duty by putting into effect the laws against fraud, misrepresentation, and obtaining money under false pretences in such cases as these electric swindles? The Patent Medicine Laws should be modified, so that every patent medicine bottle, box, or packet issued shall be required to have placed conspicuously upon it a label setting forth its actual composition. The sale of patent medicine stamps would doubtless be diminished by such a salutary regulation; but the term "patent" would then be a reality instead of, as at present, a ridiculous misnomer, while the loss to the revenue would be a trifle as compared with the saving of public money, and even of life.

[This article was published originally some four or five years ago, and it led to the downfall of one of the most gigantic advertising humbugs of the day. Respectable papers refused to insert the advertisements they had previously admitted into their columns; dupes began to see their folly; and, the better portion of the London press joined us in exposing the extensive system of deception carried on under the high-sounding title of the Electropathic Institute.]

Chapter VI.

St. Jacob's Oil; Mother Seigel's Syrup; Mattei's "Electricities."

In the last chapter we suggested certain desirable—indeed, urgently needed—modifications of the Patent Medicine Law.

It would seem that at least one patent medicine proprietor desires a change from the existing misleading system. Turning over a country newspaper lately, we noticed, artfully mixed in with items of news, a curiously-worded advertisement of a particular oil. It was headed "A Difficult Case," and had at the top of it a wood-cut illustration of a judge seated on the bench, and looking very grave and perplexed. The advertisement is couched in the anecdotal style now much affected by patent medicine vendors, and commences thus:—" There is a certain learned judge who sits in one of the London High Courts of Justice, who says that patent medicines, or —what he is pleased to term them—' quack medicines,' should never be used, except on the advice of a medical man. The judge may be correct as to the large majority of patent medicines, but there are exceptions to this rule." "*Pleased* to term them!"

The subject is not a pleasing one, at any time, but if a judge in court was pleased to call patent medicines by any other name than quack medicines he would sacrifice his reputation for truth and discernment. The exception to the rule is, according to the advertisement, St. Jacob's Oil; and details are given of the case of a young man living at a village near Uppingham, with the avowed intention of proving this assertion.

The narrative says that the symptoms were of a rheumatic character; that the patient was under the treatment of the family doctor from January, 1883, to October, 1886, when he gave the patient up as beyond his skill, and ordered him to be sent to the Leicester Infirmary; that, at the end of two weeks, the patient was discharged from the Infirmary as incurable; and that, as a last resource, he was placed under the care of a celebrated physician in Leicester, where he remained for a long time, but continued to get worse. Finally, his case having been pronounced incurable by "some of the most celebrated medical men in the Midland counties," a few bottles of St. Jacob's Oil were bought at the instigation of a neighbour, and after applying the contents, the patient was able to get about on crutches, becoming perfectly cured by January, 1888. The cure is spoken of as "miraculous," but we do not attach much importance to this expression, as it commonly occurs in patent medicine advertisements.

Indeed, we are not sure that we should have given a second thought to the matter, but for the circumstance that the advertiser complains that the preparation in question is "under the ban of a patent medicine." What does he mean? We presume that he objects either to the alternate name "quack medicine," or to the judge's opinion that, being such, it ought only to be used on the advice of a qualified medical practitioner. If the judge's ruling were followed, the manufacture of St. Jacob's Oil would sink at once into insignificance, for any medical man would be most culpably indifferent to his patient's interests, as well as to his own professional character, if he prescribed St. Jacob's Oil, seeing that this marvellous, miracle-working preparation is made of common ingredients possessing no remarkable curative properties, as will be shown presently.

Having a better acquaintance with medical men in the Midland counties, and elsewhere, than the proprietor of a quack nostrum could be expected to possess, we were curious concerning the "most celebrated" who had, in sporting parlance, thrown up the sponge after futile struggles with the mysterious ailment, so marvellously cured, and we asked a friend to write to the patient's father, inquiring their names. The reply, which we give exactly as written, was as follows:—

"In ancey to yours of this morning which we recevd, you ask me to state the case of my son the treatcmt he receved at Leicester was at the infermery and from an old Docker which has passed a way since then, and then he" (the son we, imagine, not the 'Docker') "was sent to the Devenshire Hospital Buxton and from their he was sent home quite a cripple on cruthes, and was al formes and then i gat the oil and i have the same greate faith in them as i always have done."

It is a singular misfortune that, like "the Docker" referred to in the foregoing epistle, the doctors whose professional reputations are so mercilessly marred in quack testimonials die before the testimonials have wide circulation; but, as we mentioned in EXPOSURES OF QUACKERY, Vol. I., page 90, it was a misfortune to the proprietors of Mother Seigel's Syrup that they assumed that a certain doctor had passed away in one sense, when he had only done so in another, for he had merely changed his residence. Their mistake about the "*late*" Dr. Dacre Fox cost them several thousand pounds.

It is also a remarkable circumstance that the people from whom patent medicine proprietors get such unstinted praise are usually persons—we will not use the hackneyed phrase, "whose education has been neglected"—who have as obvious a disregard for grammar and literary correctness as the patent medicine proprietors themselves have for veracity. If the case were otherwise, we might get more coherent and reliable statements of facts, and thus have a proper opportunity of forming definite opinions.

To return to the question of the composition of St. Jacob's Oil. What does this miracle-working

St. Jacob's Oil really consist of? Some, perhaps, will conjecture that the costliest, the choicest, the most delicate essences must be united to constitute so marvellous a remedy. But if so, then Dr. Selkirk Jones must surely have made some mistake. He writes:—

> In accordance with instructions, I have purchased a bottle of St. Jacob's Oil from a chemist here, and have submitted the same to a careful qualitative analysis. I find the contents chiefly comprise oil of turpentine, in which is dissolved ordinary camphor, and scented with an essential oil (most probably oil of thyme).
>
> As a medical man, I am of opinion that this Oil possesses no special therapeutic value, but, as in the case of ordinary embrocation (such as the Compound Camphor Liniment of the British Pharmacopœia) the relief afforded to the patient by its use is derived principally from the mode of its mechanical application, whereby rubefaction is produced and blood circulation accelerated. Indeed, this Oil may be regarded as an ordinary stimulating application and *nothing more.*
>
> GEORGE SELKIRK JONES, Ph.D., L.S.A., &c.

According to the Old Testament, Jacob was badly used by his future father-in-law, but the treatment he received at Laban's hands was trifling as compared with the indignity heaped upon him (after dubbing him Saint, too) by appropriating his name to a mixture of common vulgar "turps," crude camphor, and oil of thyme, and adding insult to injury by claiming for such stuff that it possesses miraculous powers!

"Turps" do cost money, though, as any housepainter would tell us, so that St. Jacob's Oil may be regarded as of higher commercial value than Mattei's Electricities, which are so miraculous in their action, according to some of Mattei's dupes, that, when administered in doses of a few drops, they will cure cataract, and unite broken bones! Mr. Stokes' analysis showed that the so-called "electricities" were water—"nothing more," as the old song says. We have been loaded with anonymous abuse and threatened with legal consequences on account of our comments on Mattei's watery wares; but, somehow, as in other instances of our publishing analyses of quack medicines, our anonymous assailants turned out to be of the Bob Acres' stamp. Mr. Stokes, too, stood firm as to the accuracy of his analysis. The following analysis and report furnished by Professor Michaud, chief of the Cantonal Laboratory of the Department of Justice at Geneva, bears out to the very letter all that Mr. Stokes reported:—

Analysis of five phials of Mattei's Electricities obtained from Mattei's Depository, June 9th, 1892.

The examination of the five different phials has given the following results:—

> About fifteen grammes of a colourless liquid, without odour or flavour.
>
> Chemical composition *indentical* with **that of** *pure water*.
>
> Contains deposits similar to those of stagnant water tainted.

The microscopic and physiological examination confirms the chemical analysis as well as a previous analysis made by Professor Stokes, of London, who declared the *therapeutic value* to be *negative*.

Mr. Stokes' conclusions seem to me to be *conclusive*.

L. MICHAUD, Professor,
Cantonal Expert in Chemistry.

Chapter VII.

OUR CORRESPONDENTS AND CRITICS; SILVERTON'S REMEDIES FOR DEAFNESS; UNQUALIFIED PRACTITIONERS: "A MERCIFUL MEDICINE, MORE PRECIOUS THAN RUBIES!"

"Man is a dupable animal. Quacks in medicine know this, and act upon that knowledge."—SOUTHEY

"QUACKS in medicine" have had their feelings considerably disturbed by the revelations contained in this series of articles, if we may judge by the abusive letters, occasionally varied by threats of action for libel, that have reached us.

As regards the former, always anonymous, the attention we have bestowed on them has extended as far as our waste-paper basket; while the threats have been as vague and incomplete as the analyses published have been positive and definite. Whenever our angry correspondents think fit to put their threats into tangible, business-like form, we shall be prepared to prove up to the hilt every statement we have made; but we fancy that they are not so courageous, or so simple-minded, as Oliver Twist was when he asked for more.

In fact, some of the patent medicine people seem to have had too much already. One patent medicine company, to whose remedies (!) we gave gratuitous, though world-wide, publicity, has lately held the annual meetings of its shareholders, and the directors have had to explain that the great falling off of receipts, and consequent shortness of dividends, are due to the influence of efforts used to stop the sales of their "marvellous medicines;" while, as regards another company, whose sheet-anchor is, as our published analysis proved, only saltpetre, the *Financial Gazette*, of a recent date, says of it that the shares have dropped so low as to be "practically unsaleable."

Whatever the views of patent medicine makers may be, there is, at any rate, a remarkable consensus of opinion amongst the Press, and we are constantly receiving papers containing most favourable, indeed flattering, notices of " Patent *alias* Quack Medicines." We regret that our limited space prevents our acknowledging all of these; let one suffice as a specimen. Our excellent contemporary, the *British Medical Journal*, of October 1st, speaking of our journal, says :—" Its editor has for some time past devoted himself to the task of exposing the later forms of medical quackery; and from its interesting articles we learn all that the analyst has to tell us about these loudly-trumpeted preparations." The reviewer concludes as follows :— " As a contribution to our knowledge of the ways of quacks, it is to be welcomed; and we hope that all of our readers will make it their business to acquaint themselves with its contents."

Letters referring to our articles, from correspondents living in Great Britain and abroad, have been most numerous. Here, also, we must ask to be allowed to give only a single specimen. An Indian judge writes:—"I have just finished reading, with great amusement and profit, your delightfully humorous and powerful articles. I am glad to find that your exposures of quackery continue to appear in HEALTH NEWS, and I congratulate you upon the good work that you are doing." Many of our correspondents are medical practitioners, who inform us that they have succeeded in promptly and effectively opening the eyes of believers in this or that nostrum by putting in their hands the articles bearing on it.

Our quotation from Southey, reminds us of a curiously-worded card which was put into our hands some time ago in Fleet Street, near Ludgate Circus. It ran thus:—"Admit the bearer to a free consultation on deafness and noises in the head and ears, with the London specialist, the Rev. E. J. Silverton, now returned from a long tour through Glasgow and all the large towns of Scotland; Liverpool and all the large towns of England; also Ireland and Wales, where wonderful cures have been performed without operation or the use of instruments. The treatment is so gentle in its action, that little children are often saved from being Deaf and Dumb; and some who were deaf and dumb have been made to hear, and then taught to talk. Old standing cases are also successful"—the language is as mysterious as the gentle method—" indeed, it seems no more difficult to cure at seventy-two than at seventeen. The patient should not be discouraged because he or she has tried before and failed. Mr. Silverton has been in the *work* over twenty years" this expression smacks rather of the parson than of the physician — " and has been eminently successful "—just now the cases were successful—" in all kinds of cases"; here followeth, to employ the clerical phraseology, a long list of ailments, which we need not inflict upon our readers. "If the case is incurable no hope will be held out, but valuable advice will be given to each sufferer" —" *given* " has a philanthropic sound about it, but there the philanthropy ends,—" and where there seems to be hope, remedies will be recommended."

"CAUTION"—we must put this word into capitals, as on the card lying before us, although we could not spare capitals enough for all the words thus printed. "Caution," we say, in more senses than one, knowing what we do; but we will give the " reverend specialist's " language, not our own. "Patients may bring one friend, but we have not room for four or five people to come with one."

It seems a singular thing that four or five people should even wish to accompany, much more that they should, unless strictly prohibited, accompany every individual patient to a consultation. We have been in practice twice as long as the "reverend specialist" has been "in the work," and we can only remember a single instance of four persons entering our consulting-room with one patient. On this exceptional occasion, the patient was suspected to be of unsound mind, and the quartet accompanying him, being composed of near relations, may possibly have regarded it a good opportunity for getting *(at his expense)*, an indirect tip as to their own mental condition. But, surely, the "reverend specialist's" visitors are not all afflicted with weak intellect, though many might be suspected of a failing in that direction; rather let us suppose that they are actuated by a burning desire to see him "on the job"—"in the work," we mean.

However, those who are shut out by the "reverend specialist's" imperative order need not, like Moore's Peri, in "Lalla Rookh," "at the gate of Heaven sit disconsolate." The "reverend specialist" has made arrangements for their supply with his patent pills and other patent remedies, which, similarly to his treatment of deafness, would appear to be equally well suited to the patient's case, whether seventeen or seventy-two years of age, or of either sex.

At any rate, we have our reasons for this supposition. We happen to know that some years ago, two people, a gentleman and a lady, consulted the "reverend specialist," in the same week, in consequence of seeing his advertisements in the newspapers;

the gentleman seeking advice for rheumatism, the lady for sterility of many years' duration, a condition which she was anxious to have remedied, if possible. In accordance with the "reverend specialist's" rule that, "where there seems to be hope, remedies will be recommended," the lady received medicine as well as the gentleman. In the gentleman's case thirty-five shillings were demanded for what would be dear — commercially speaking — at thirty-five farthings. The pills and mixture supplied to each were examined by a public analyst, and in each case the remedies were identical in composition; the mixture being composed of water, chloride of iron, acetate of ammonia, and syrup of orange to flavour and colour it, while the pills were merely common rhubarb pills. Yet these very ordinary pills were described by the "reverend specialist" in one of his pamphlets as "a merciful medicine, more precious than rubies"! Presuming that this is a fair sample of what people get when they visit the "reverend specialist's" consulting rooms, where he can be seen, with "his physician in attendance," it would be a difficult question to decide which is the more simple, the patient or the treatment.

The object of the "reverend specialist" in associating with himself a medical man holding some qualification or other is obvious—namely, to evade the law relating to unqualified medical practitioners. But if the Medical Council, whose duty it is to protect alike the interests of the profession and the public, would exercise the power vested in them to its full extent, such a state of things could not exist as that of an unqualified man, boldly announcing himself to be possessed of special knowledge, far beyond that of others who have devoted their lifetime to medical study, and as boldly practising, under the cover of an alleged doctor, whose very name is suppressed. Judges, coroners, and magistrates are properly severe when cases come before them in which unqualified medical assistants have attended patients; is not the case of a qualified medical man with an unqualified associate as his employer still more deserving of reprobation and punishment?

As a matter of fact, the Medical Council have it in their power to deal with cases of this nature, under the Medical Act of 1858. Indeed, they sometimes exercise this power, and not very long ago two practitioners were struck off the Medical Register, the charges proved against them being that they had carried on practice in conjunction with, or acted as cover for, unqualified persons.

In a pamphlet largely circulated by the "reverend specialist," it is stated that, prior to commencing his present career, he was a Baptist minister. But, "pressure of pastoral duties and the very wide range of his healing ministry"—wider, apparently, than his range of drugs, or acquaintance with the properties of those he did use—"compelled him to retire from either the one or the other. Mr. Silverton has retired from the pastor's office, to the great regret of many hundreds of warmly attached friends constituting the Church meeting in Exeter Hall, Nottingham. It seems as sacred to give a man health as to bless him religiously; but often the one leads to the other." As the "reverend specialist" probably wrote this modest eugolium himself, we will content ourselves with expressing the sincere hope that not so large a number as the " many hundreds of warmly attached friends" have had reason to regret his change of vocation, though we confess to a good deal of doubt on this point.

We are informed that the building at Nottingham bearing the high-sounding title of Exeter Hall, is only a small chapel; also that the "reverend specialist" used to quack while actively engaged in raising funds for it;—or, as he perhaps put it, was receiving the worldly dross derived from the cure of bodies, while devoting himself to the spiritual cure of souls.

Chapter VIII.

Patent Medicines and Pious Language; the "Reverend Specialist"; Congreve's Balsamic Elixir; Owbridge's Lung Tonic; Lane's Catarrh Cure; a Quack's Certificate.

"*'Tis not the many oaths that make the truth.*"—Shakespeare.

Many of our press contemporaries are in the habit of stimulating the patronage and exciting the hopes of their readers, by offering prizes for essays on various subjects, or for the solution of different conundrums. We are happy to say that the demand for Health News is too genuine to require any incentive of this kind. But if we thought that the case was otherwise, we might feel tempted to hold out the inducement of a ten-guinea prize, to be competed for by believers in patent medicines only (namely, ten boxes of Beecham's Pills*), and the subject of the conundrum-essay would be, "How do you account for the very frequent relation between patent medicines and pious language?"

---

* "Worth a guinea a box" Beecham's assertion. "Worth a penny a box," the real fact. See Vol. I. of Exposures of Quackery, page 93.

In our last number we quoted from, probably, Silverton's own words, a description of the desperate conflict in a "reverend specialist's" mind between his devotion to his pastoral work, and a presumed call to medical duties. [N.B.—We are not referring here to the three halfpenny stamps for Government duty on each box of quack pills.] If it were not bordering on profanity, in presence of such a noble struggle, we might have compared the "reverend specialist" with Garrick, as depicted in the famous picture of that actor hesitating whether to adopt tragedy or comedy as his sovereign queen.

We may here explain that the "reverend specialist" got over his difficulty by arriving at the conclusion expressed in the following unctuous sentence :—" It seems as sacred to give a man health as to bless him religiously; but often the one leads to the other." The "but" somehow mars the force and the intelligibility of the sentence. Blessing religiously, too, reminds us of an anecdote of two sailors on board a ship which included a colonial bishop amongst its passengers. The bishop objected to the emphatic manner in which the simple sons of Neptune are wont to express passing sentiments; and the captain impressed upon the crew the desirability of controlling their feelings, or at least their language, during *that* voyage. The sailors were obedient to orders, and almost mute throughout the early part of the day, owing to the novelty of the restraint put on them;

but in the course of the afternoon one of the men happened to drop a heavy marling-spike upon another's foot. Turning sharply round, the latter, catching sight of the captain and the bishop close by, roared out, "*Bless you, Jack!*" adding in an undertone, "You know what I mean." Evidently, the injured sailor stood in little need of one of the " energisers " which the " reverend specialist " sells for about half-a-crown to anyone willing to part with that coin of the realm; for not only does Silverton profess to cure deafness, even where all other remedial measures have failed, but he *energises* those who need it with his patent " energisers " at 2s. 6d.; whilst his rhubarb pills (1s. 1½d. a box) are, according to his modest assertion, " a merciful medicine, more precious than rubies ! "

We must now leave the " reverend specialist ; we have taken up too much valuable time already in exposing his nostrums, and, when our publisher sees this article in print, he may blame us for giving gratuitous advertisements. However, he cannot say that we have reduced the receipts in that department, as there is an inexorable rule at the Office of HEALTH NEWS as unyielding as the laws of the Medes and Persians, " Quack Advertisements Rigorously Excluded." In order that there shall be no mistake on this point, this announcement is printed on the front page of every number of the journal.

Referring further to this religious (?) tendency on the part of patent medicine proprietors, we may remark that one of the largest advertisers of that class heads the monster announcements with a quotation from Deuteronomy : "For the blood is the life." The reason for dragging in a Scriptural quotation in this manner is not very evident, except that it be for the purpose of drawing attention, just as the Sequah men draw teeth. Besides, having regard to the fact that Clarke's Blood Mixture contains a large quantity of iodide of potassium, and that this drug has a powerfully depressing action, making it most unsuitable and improper for administration to the sickly and debilitated (to whom it is, nevertheless, strongly recommended by the makers), we think that a more satisfactory quotation could have been taken from a verse which occurs later in the same chapter, "Pour it on the ground;" at any rate, this would have had a certain cautionary value, and have suggested some safer way of dealing with the Blood Mixture than swallowing it.

As for "God's blessing," patent medicine people use the expression so freely (to such an irreverent extent as to partake of blasphemy), that one is almost led to imagine that they regard it as included in the purchase when they buy the Government duty three-halfpenny stamps at Somerset House.

We have lying on our table a book which is supposed to treat of consumption and other chest diseases. At all events the names of these affections appear in gilt letters upon the cover; but on opening it we find only twenty-five pages, with frequent digressions in praise of the Balsamic Elixir, concerning these important affections, while about one hundred and fifty pages are stuffed with testimonials Yet, in what may, by a great straining of the phrase, be termed the professional portion, the author finds himself obliged to utter the ridiculous excuse for not describing asthma, that "the limits of the present work will not permit me to enter minutely into such matters." We do not believe that he could, if he tried, even if testimonials were not the easier task. And as to the testimonial portion, occupying six-sevenths of the whole book, he says, "Other extracts might be given, but want of space forbids." What a pity it is that his modesty stood so much in his way that he did not utilise the odd twenty-five pages for testimonials; there would have been no loss to science, and the book would have had just as much weight (eight ounces for twopence) when "gradually diffused" through the medium of the post, amongst its credulous readers.

On opening this book, which purports to be wholly written (with the exception, we presume, of the 150 pages of testimonials, though the statement is not very clear, and patent medicine literature must always be taken with distrust) by one George Thomas Congreve, of Peckham, we find—we felt sure we should—" God's blessing" figuring on the very first page; G. T. C. expressing the hope that "by gradually diffused knowledge"—we once heard an enterprising agent and bill-sticker described as a "professor of *applied* literature;" is this the sort of gradual diffusion meant?—"a just appreciation of these means and their principles of operation will so arise that, with God's blessing, the benefits accruing therefrom may be extended into distant lands, as well as more widely in our own." Opinions differ; consequently, there are doubtless some who will entertain opposite views to Mr. Congreve, both as to the "principles of operation" and the desirability of invoking the Divine benediction upon them.

The preface to Mr. Congreve's book of testimonials is amusingly contradictory. He is at great pains to explain that for many years his attention has been "earnestly directed" to the study of pulmonary disease; first with his father, then as a pupil of the "late J. R. Hancorn" (whom it is to be hoped he treated with a little more respect when speaking of him in his life-time than he does now in writing of him), and afterwards in the medical schools and hospitals of London, where he "attended all the courses of lectures and clinical practice required of the medical student." The medical schools and hospitals of London are numerous, and we cannot help thinking that it would have facilitated Mr. Congreve's studies had he applied himself steadily to work at one institution, instead of imitating the bee, roving from place to place in quest of honey—Balsamic Elixir, we should perhaps have said.

A country squire, whose son used, like Mr. Congreve, to boast in the plural number of the seats of learning with which he was acquainted, was descanting upon the subject to one of his tenants, and commented upon the remarkable circumstance that the young Hopeful had enjoyed the exceptional advantage of being educated at two universities. "There isn't much in that," replied the old farmer, who had not the high opinion of the young squire's mental abilities that the father had; "why, I once had a calf that sucked two cows, and the more he sucked the bigger calf he grew."

Well, supposing that Mr. Congreve's father was a medical practitioner, although by the way, Mr. Congreve does not furnish corroboration on this point; that J. R. Hancorn (it goes against the grain to have to speak thus familiarly of the eminent surgeon who had the inestimable honour of imparting the principles of special knowledge to Mr. Congreve) "had an extensive practice in cases of consumption," as Mr. Congreve asserts; and that Mr. Congreve attended all the hospitals and medical schools of London, and all the courses of lectures required of the medical student, there is a certain something which should be explained, and that is, why he omitted what is required, or expected, of every student—namely, to pass an examination at one or more of the colleges, thus affording evidence of proficiency, and qualifying for medical practice. Perhaps, despite the all-absorbing character of his medical studies, he stole an occasional half-hour for the perusal of Shakespeare, and the following passage in *King John* struck his fancy as appropriate to his case :—

> Therefore, to be possess'd with double pomp,
> To guard a title that was rich before,
> To gild refined gold, to paint the lily,
> To throw a perfume on the violet,
> To smooth the ice, or add another hue
> Unto the rainbow, or with taper light
> To seek the beateous eye of heaven to garnish,
> Is wasteful and ridiculous excess.

Whether such a supposition is correct, or not, cannot be determined here; but one thing is a fact, namely, that Mr. Congreve does not think it expedient to wholly disregard professional qualifications, for he tells us, in imposing capital letters," I have much pleasure to announce that I have secured the valuable assistance of my son-in-law, J. Alex. Brown, M.R.C.S., L.S.A., who has now been with me some years. In my absence patients will be carefully attended to by him." Anyone with a little imaginative power might think he was reading of a modern Laban and a second Jacob. But Laban restricted Jacob to looking after sheep, while Mr. Congreve entrusts his son-in-law with the charge of patients in his (Mr. Congreve's) absence. What, we wonder, do the patients, if they share Mr. Congreve's lofty contempt for medical qualifications, say to this "wasteful and ridiculous excess?" And what ought the Medical Council to say to such an extraordinary arrangement?

The "reverend specialist," referred to in our previous article, employed " a physician in attendance" obviously as "a cover;" but if this be Mr. Congreve's motive for securing the valuable assistance of his son-in-law, it could not avail him much in the event of proceedings being taken for illegal practice; for he naively admits that J. Alex. Brown, M.R.C.S., L.S.A., is permitted to see Mr. Congreve's patients only when Mr. Congreve is absent; to use a phrase common amongst boys sliding on the ice, J. Alex. Brown's sole function and privilege appear to be to "keep the pot a-boiling."

As Mr. Congreve reminds his readers in the cheerful (?) Christian style which pervades the entire book, there is "an appointed time for man on earth," and if anyone wishes to personally obtain the "*God-provided* remedy for poor suffering humanity" (the Balsamic Elixir is thus beautifully and modestly (!) described at page 87), he must present himself at Mr. Congreve's residence on certain days at stated hours, when he can consult Mr. Congreve, or, in his absence, J. Alex. Brown.

Some anxious inquirer may despondingly suggest such a misfortune as the absence of both of these individuals. We hasten to dispel the gathering gloom which must result from even the bare idea of such a calamity. "In order that the world at large might derive the benefit of their use," Mr. Congreve tells us that these "God-provided remedies," "prepared solely by me, at my residence," &c.—that accounts for the days on which he cannot be seen, we suppose—have been " introduced in the form of *proprietary* medicines," which can be had of any respectable chemist.

Imaginary dialogue in a village shop between a chemist and a lady-customer:—" Are you a *respectable* chemist?" "Why, certainly." "Then I want a bottle of Mr. Congreve's '*God-provided*' Balsamic Elixir." "Yes, madam; small or large size? Mr. Congreve, the sole providee (I mean preparer) says: 'The circumstances of the patient permitting, it is much better to have the latter—a saving of trouble and expense.' "What is the price?" "Family bottles are 11s. and 22s." "Give me a twenty-two shilling bottle; it is so much cheaper than my paying a few shillings to a qualified medical man who, of course —at least, so Mr. Congreve says in his book—would not know anything about my case."

Mr. Congreve professes not to mind what trouble he takes. "I am not actuated solely by ideas of pecuniary gain, but a sincere desire to benefit my fellow-creatures as much as my own advantage," he tells us in his book. Nay, more, he does not mind apparently, how much trouble he is put to, in addition to solely preparing the remedies, for he specially remarks—using italics to emphasise the fact of his earnestness—"*whenever the two largest sizes are required, it is better to obtain them direct from me.*" This admonition is needed, for otherwise the thoughtless patient might imagine it would be a saving of trouble and expense to buy of some chemist—respectable chemist, of course—nearer home, Why there is one not far from our office who actually sells the 11s. size for 8s. 9d. and the 22s. family bottle for 17s. 6d., as we learn from his price list. We have never made such wholesale purchases of any patent medicines. Life is not long enough for such experiments, let alone our purse; but if we ever found ourselves so silly as to indulge in such "wasteful and ridiculous excess," as regards the Balsamic Elixir, we should probably take the first omnibus to Peckham, carefully sitting back in a corner as we passed the lunatic asylum (lest our object might be guessed, and ourselves be detained as fitting candidates for admission), and save trouble and expense (?) by paying the full price into the hands that had solely prepared this *God-provided* nectar. It is so much more satisfactory to get a bottle of wine direct from the original bin than to have it from the public-house round the corner.

Now, what on earth could have put this last idea into our mind? Why, we read some time ago in the *Provincial Medical Journal* an article in which the Balsamic Elixir was irreverently compared to diluted "publican's port," *plus* a little Friar's Balsam, also known as the compound tincture of benzoin. Such a concoction could not, by the greatest stretch of language, be rightly termed *God-provided*, and we have therefore had the stuff recently analysed. The analyst to whom we submitted the sample, a gentleman of many years' experience, reports thus:—" It seems to be made of infusion of elderberries, a little alcohol and benzoic acid, with a little allspice to flavour it." There is an old Scotch proverb that "Many a little makes a mickle." If so, the "mickle" in this case is all on the side of the sole preparer, and not in favour of the consumer. Country-born and country-bred, we admit an inherant weakness on a cold winter's night for mulled elderberry wine, *plus* sugar, *plus* spice, but we should strenuously object to its being chemically "mulled" by the addition of compound tincture of benzoin, *alias* Friar's Balsam.

As for the properties of *this* beverage—we are now speaking of good, honest, unsophisticated elderberry wine—they are too well known for us to need to describe them; but as for the remedial properties of *that other* concoction—Congreve's Balsamic Elixir, to wit—we most unhesitatingly and positively assert that it has no power whatever to prevent or modify tubercular deposits, to heal ulcerated lung-tissue, or to cope with the various pathological conditions giving rise to asthma and other chest affections, alleged to be cured by the administration of the Balsamic Elixir; no, not even if such were affirmed in twenty times the number of affidavits and testimonials put forward by Mr. Congreve. " 'Tis not the many oaths that make the truth," as Shakespeare observed.

This chapter has already run to such a length that we cannot make room for any extended comments on certain much advertised proprietary medicines which we had intended to write about. One of these is Owbridge's Lung Tonic; a second is "Dr." Lane's Catarrh Cure, sold at "Professor" Brown's depôt for herbal medicines, in a street near Covent Garden. Frequent newspaper advertisements exhort people to "save their lives by taking Owbridge's Lung Tonic;" and the proprietors of the Catarrh Cure insist upon the statement that its use as a gargle will prevent consumption. How far the facts will agree with the assertion may be readily realised when we mention that the tell-tale test-tube of our analyst makes us acquainted with the fact that the simple solution of a drachm of carbolic acid in a pint of water (*both* cheap ingredients) furnishes a compound closely proximate to the contents of a 4s. 6d. bottle of the infallible herbal (!) Catarrh Cure and Consumption-preventer.

The other life-preserver, Owbridge's Lung Tonic, turns out to be composed chiefly of balsam of tolu, with the oils of aniseed and of cloves. Aromatic, warming, and not unpleasant to the palate, but possessing one quality in common with the two preparations already described—the Balsamic Elixir and the Catarrh Cure—namely, utter uselessness in the prevention or cure of the fell disease, consumption, for which the makers falsely assert that they are specifics.

A clergyman in one of the Eastern counties has sent us a copy of a travelling-van quack's certificate, given to a labouring man for the purpose of obtaining sick relief from a benefit club:—

This is to certefy that James——is sufering from yaler janders and not abel to work.
G. LEWES, M.D., U.S.A.

The originality of the spelling is sufficiently striking; but there is another peculiarity about the certificate, and that is, that the U was scrawled, with a long top stroke, so as to resemble the letter L; the obvious purpose being to convey to ignorant or unsuspecting persons that, besides the doubtlessly assumed degree of M.D., the writer of the certificate possessed the L.S.A. qualification. Our subscriber, happening to see the certificate, made some inquiries, which had the effect of promptly relieving his parish of an impostor. What a pity it is that there are not more clerical gentlemen ready, like this one, to oppose quackery, instead of many being so easily led into giving testimonials about matters of which they probably know nothing at all.

Chapter IX.

Stepney Green Diplomas.

In a former article, we spoke of the so-called " School of Safe Medicine." A greater peice of humbug never existed. Four men banded themselves together, took a small house in Stepney, dubbed it "The School of Botanic and Magnetic Medicine," and themselves "Professors," and forthwith proceeded to issue diplomas and degrees to anyone who would pay them for these worthless documents. Such a queer quartet, too! Here they are: Charles Gapp, not many years ago a coal-dealer's clerk, next a cheap jewellery vendor, and now carrying on a "medical practice" at the East End of London, and putting "M.D." after his name, on the strength of a bogus degree of the United States, which he obtained without even leaving England; "Dr." Younger, of the Alofas* *(all-a-farce)* shop in Oxford Street, London, near Holloway's Temple of Quackery, holding a similar diploma to Gapp; Maguire, a "botanic" medical practitioner at Bow, who acts as secretary; and the "Reverend" Verryman Trimmings, whose M.A. and Ph.D. emenate from the same tainted

---

* See Exposures of Quackery, vol. 1, page 99: "The Alofas *(all-a-farce)* Safe Remedies."

source as Gapp's and Younger's M.D. According to a most reliable authority, the last-named may lay claim to be considered a "man of letters" and artistic culture, seeing that at a certain period—not very remote—before be became a "reverend," and, like Gapp and Younger, *took* his degree, Verryman Trimmings was actively engaged in the sale of a peculiar class of French literature and photographs, in a small shop in Holywell Street, Strand.

The system of education pursued at this school or college—its four shining lights appear undeterminded which to call the establishment—is as simple as the poor dupes who are afterwards gulled into believing the degrees to be of any value. Postal tuition in medicine, surgery, magnetism, mesmerism, botany, etc., postal examination (we should like to see some of the questions and answers), postal graduation—everything postal, provided, of course, that the postally posted candidates "post" the £25, a point on which the four professional examiners are very stringent.

These Stepney Diplomas are not, as a matter of fact, worth twenty-five pence, let alone twenty-five pounds, except to hang in some herbalist's back shop, pretentiously designated "the consulting room," so as to assist in conjuring hard-earned money out of credulous customer's pockets.

In 1894, the queer quack quartet, *alias* the School of Safe medicine, aided one of their "graduates," Joseph Steel, of Houghton-le-Spring, Durham, described as a "miner and herbalist," in appealing against his conviction by the magistrates, who imposed a fine of £10 with £6 costs, for falsely passing himself off as a doctor, though holding only a Stepney Green diploma, which, we believe, he received without the trouble of coming to London. Our readers will readily guess the result of the appeal, though it does not seem to have abashed the impudence or damped the ardour of the Stepney Syndicate for the issue of sham diplomas, as they have since blossomed into the "General Council of Safe Medicine," thus named in obvious imitation of the General Medical Council, and are energetically engaged in soliciting assistance to enable them to establish the Magnetic and Botanic School of Safe Medicine, Limited.

Until they have succeeded in raising money to erect a Magnetic and Botanic Hospital, the General Safe Council graciously express willingness to accept the certificates of attendance granted by "any recognised medical school"; so that St. Bartholomew's, Guy's, St. George's, the London, and other *at present* recognised medical schools, may possibly hold out a little longer.

The General Safe Council announced that communications and donations might be sent to the Secretary (Verryman Trimmings again to the front), at the offices of the General Council of Safe Medicine, Limited, Club Union Hall, next to Holborn Town Hall, London. In noticing this appeal a little time back, we also advised people to *send* their communications, if any were felt desirable (it will be seen that we did not say anything as to donations, leaving that matter to their common sense), for our own inspection had shown us that a personal application would be attended with risk in working one's way down into the underground room used for the several purposes of the General Safe Council, the Botanic and Magnetic Dispensary, the massage and mesmeric mummeries, &c. The address, as printed and circulated, would almost have led anyone not acquainted with the place to suppose that the General Safe Council occupied the entire building, and might soon requisition the Town Hall. But the General Safe Council were as chary of their outlay on rent as they were niggard in the time which they gave to the "philanthropic work," and from what several indignant members of the club (who did not appreciate the close proximity of the Council) told us it was evident that before Gapp, Younger & Co. absorbed the Union Club Hall, many hundreds of members, as well as many other tenants of the premises, would, like Macaulay's Cornishmen, want to "know the reason why," particularly as even the limited underground apartment was only used once a week, viz., on Thursdays, at 8 p.m., when the General Safe Council met, when the magnetic and botanic dispensary threw open its portals to "thousands of sufferers"—where they could put them all is a mystery that nobody could solve—and when the quartet displayed their vaunted skill in massage, mesmerism, and animal magnetism, principally to the weaker sex.

Whether our critical comments on the Council's arrangements affected the sensitive Stepney Syndicate's nerves, or whether they pined for greater seclusion, we know not. But some months after curiosity had led us to dive down into their subterranean quarters, we made another visit to the Club Union Hall, and on inquiring of some members grouped round the entrance-door we were informed, with much show of merriment, that the "crew" had gone back to Stepney. What they are doing at Stepney is, that they are turning out batches of quack doctors, to the scandal of the Law and of the authorities whose duty it is to put down such public swindles, and to the infinite danger of the pockets, health, and lives of thousands of people deluded into believing that these diplomas are of value.

Not long since, we received a letter from a Colonial reader, telling us that a man, provided with a Stepney Green diploma, had lately found his way to Australia (not by transportation, for that antiquated mode of emigration is done away with), and was working much mischief amongst the ignorant denizens of the poorer district of the city in which our correspondent resided.

The unblushing effrontery with which the Stepney Green "graduates" parade their M.D. degrees nearer home may be judged by the sub-joined fac-simile of a label forwarded to us by a subscriber to HEALTH NEWS, resident in the North of England. He also informs us that "Doctor" Thomas Hudspith adds a sweet-stuff shop to his other "professional" attractions.

THOS. HUDSPITH,
M.D. (&c.), Dr. of Botanic Medicine,
Diploma of Incorporated Council of Safe Medicine, Ld., by Examination; also member of the Magnetic and Botanic School of Safe Medicine Ld., London, incorporated by H.M. Government in contradistinction to the Allopathic system of Drug Medication.

ORGANIC-MAGNETIC & BOTANIC DISPENSARY,
LANGLEY MOOR, Near Durham.

HERBS, ROOTS, BARKS, AND ALL BOTANIC SAFE MEDICINES.

Here is a deliberately misleading statement about H.M. Government which the authorities ought to look into. The British Government is behind almost every Continental one as regards adequate measures for protecting the public from quackery, and the same supineness appears to affect our officials when they go abroad. During the course of the trial, at the Old Bailey, of the wretchedly ignorant pretenders to surgical knowledge, who, styling themselves "Indian oculists," robbed thousands of poor people, in all parts of England, of their money and sight, it came out in evidence that some documents which these men represented to their unfortunate dupes as Indian diplomas, were passports from the Indian Government, stating that the holders were "proceeding to Europe to *cure* eye diseases!"

Chapter X.

Bonesetting, by a Hospital Surgeon; a Patent Medicine Song, By John Johnston, M.D.

In this chapter we include two contributions which have been forwarded to us, at different times; the one by a Hospital Surgeon, on a peculiar form of quackery, doubtless well-known to many of our readers in country districts, where it flourishes more than in large towns; the other, a highly humorous effusion by Dr. John Johnston, of Bolton, being a song specially composed by that gentleman for the Annual Dinner of the Bolton and District Medical Society.

BONESETTING.

A peculiar form of quackery is what is called Bonesetting. This is most popular amongst the lower ranks of society, such as the miners and dalesmen of the North of England, and small farmers, shopkeepers, and artisans in some other parts of the country, for instance; though the belief in it is by no means limited to the uneducated classes.

The process is plain in the extreme, and any ordinary blacksmith is competent to set up business in it at a moment's notice. The primitive method of simple, unintelligent violence is all the secret there is in it. A person happens to have a stiff joint, either from accident or disease, and is told by a friend of the wonderful skill of a certain bonesetter. He consequently goes to this unqualified practitioner, and is invariably informed that his surgeon has not treated his case properly; also, in nine cases out of ten, that "the small bone" of the affected joint has been put out. Perhaps the particular joint has no small bone in connection with it; but that is a mere detail, and the bonesetter, whether ignorant on this point or not, does not deem it desirable to throw away any chances of magnifying his ability; while the sudden breaking down of old adhesions, resulting from his manipulation, is, of course, attributed to the "small bone" going into its place again.

The bonesetter, having delivered his opinion in his customary manner, proceeds to bend the joint by sheer force. In some instances, when the stiffness is the result of a simple adhesion in the joint, such as is frequent after a sprain or other ordinary injury, the wrench is successful; indeed, this is a method employed in suitable cases by competent surgeons. If, however, there should be actual disease of the joint—rheumatism, scrofula, or cancer—a violent wrench by one of these ignorant men not only does no permanent good, but may set up serious injury, ending in the necessity for amputation, or even in death.

It is almost superfluous to say that this class of impostors usually claim some special Divine gift,* and it is equally unnecessary to point out that though one hears much of their occasional successes, victims of their blundering generally keep the story of pain and folly to themselves.

Mr. Howard Marsh, one of the leading surgeons of the day, in his work on Diseases of the Joints, gives an amusing and instructive account of three persons whom he sent, though they had nothing the matter with them, to a notorious bonesetter. The first was informed on showing his elbow that his ulna (the long, slender bone of the forearm) was "out," and, having paid half-a-guinea, was directed to return in a couple of days with two guineas, when the bone would be put into its place. A second was told that his ankle-bone was out, and the same course was pursued as regards fee. The third received identically the same opinion and instructions as the second.

The writer was once the subject of considerable abuse and misrepresentation by a lady, whose servant had sustained a fracture at the middle of her leg; the necessity of keeping the injured limb in splints for about a month had resulted in a temporary stiffness of the ankle-joint, such as is not an unusual occurrence in cases of this kind. The employer, not satisfied with the treatment which her servant had had in the

---

* One whom we recollect in Warwickshire used to boast that he was the seventh son of a seventh son.—ED.

hospital, took her out of that institution before she had completely recovered, and sent her to a bonesetter. Without the least hesitation this man asserted that the leg had not been broken at all, but that the small bone of the ankle was "out." As a matter of fact, there is, properly speaking, no small bone of the ankle; but the bonesetter was above such trivial details, at any rate he professed to have put it in its right place, to the intense satisfaction of the lady, who handed him a good fee in recognition of his *special* skill.

A PATENT-MEDICINE SONG.

COME, friends and brother Medicals, and listen to my song,
Though it consists of verses ten, it won't detain you long;
It's all about the marvellous notions, lotions, draughts and pills,
That are guaranteed to cure the human race of all its ills.

Of weakness of the muscles or the nerves, wherever felt,
You'll speedily be cured by wearing an "Electric Belt."
What matter if it's only made of little bits of tin?
It's called *Electric* and the metal's nicely quilted in.

For heat spots, pimples, boils, and all "disorders of the blood,"
Clarke's mixture, with its *Pot: Iod:*, can't fail to do you good,
While Mother Seigel's Syrup, with its treacle and its aloes,
Is a priceless remedy for all, from slum to Royal palace.

And should your stomach be upset, or your liver be at fault,
The thing that's sure to put you right is a dose of Eno's Salt;
'Tis true a Seidlitz Powder would have much the same effect,
But as that bears no *patent* stamp, what good can you expect?

For Rheumatism nothing can excel St. Jacob's Oil,
With its camphor and its turpentine, pure products of the soil;
For sciatica that's chronic, or lumbago in the back.
Get Sequah's Indian chiefs to rub you till you're blue and black.

That women folks are fond of pills, old Holloway could teach 'em,
But nowadays they're more inclined to pin their faith to Beecham,
Whose pills they take by handfuls with a confidence nothing shocks,
For don't they know that "Beecham's Pills are worth a guinea a box?"

For crying babes and children we have nostrums by the score,
There are "teething powders," "soothing syrups," and "mothers' friends" galore;
And while it's true that all such owe their power to "sleeping stuff,"
They soothe and quiet the little dears—and isn't that enough?

And should your hair evince a strong desire from you to part,
At once apply the lotion made by Mrs. Allen's art,
And on each bald and barren spot 'twill soon sprout up anew,
While silvery locks will speedily regain their youthful hue.

But time would fail to speak of all the wondrous things we hear,
And we marvel at the statements that in circulars appear,
How Warner's cure for instance can *cure* anything at all,
If it's true that it contains a large amount of alcohol.

In fact, unless you want to die, there seems no room for doubt,
That you must swallow every patent medicine that comes out;
And should you find by doing so you've quite destroyed your health,
You'll know at least that you've increased the medicine vendor's wealth.

Chapter XI.

QUACK BOGUS NEWSPAPERS; HOW THE POOR ARE SWINDLED; HANDYSIDE'S CONSUMPTION AND CANCER CURE; ELECTRIC SNUFF

"*How this world is given to lying!*"—SHAKESPEARE.

KING David is recorded to have said in his haste that "all men are liars." Commentators, who are not usually given to unanimity of opinion, are fully agreed upon one point, namely, that this is rather too sweeping an assertion, and needs some little qualification. For instance, if the words "quack medicine" were inserted after "all," every possible doubt as to King David's complete correctness would at once be removed.

In connection with our articles on quackery, it has been part of our arduous task to go through a large amount of quack literature, and we have invariably found in all quack pamphlets, bills, and advertisements the same utter disregard of truthfulness, prompted by the same reckless greed for their dupes' money. "Horrors on horrors' heads accumulate, the worst of evils fastest propagate"; and not content with flooding the country with thousands of tons of printers' ink annually, in the form that we have mentioned, quacks have of late years hit upon the disreputable dodge of bringing out sham newspapers, as new snares to catch the unwary and the ignorant. These never get beyond one number; it is not intended that they should, for having done the trick under one name, the title is altered, the heading lines are changed, and the same letter-press does service over again. Millions of copies of various bogus journals, puffing this or that nostrum, are circulated by post through whole counties, or distributed by hand from house to house in populous towns; always at a distance from the assumed place of origin, be it observed, though a closer investigation would show that they emanate from the same printing office. In this manner, the "Jersey Juggler" is easily transformed into the "Belfast Boaster," and so the game proceeds. The "mostly fools" who, according to Carlyle, constitute the greater proportion of the inhabitants of Great Britain, swallow the disguised bait, and take for gospel every word they read, while the charlatans chuckle over their successful swindle.

Sometimes these catchgull papers are at first sight harmless, but there is in such cases a supplement, the circulation of which is, for obvious reasons, the chief aim of the proprietor. One of these, which was sent to us by a Newcastle subscriber to HEALTH NEWS, is styled the "Supplement to Handyside's Monthly Journal; or, How to Make Money Easy, and How to Preserve Health to Enjoy it." This is an attractive, though an involved and ungrammatical, title; if Handyside had pretended to tell his readers how to make money *easily*, for instance, he would have been somewhat more intelligible. But Handyside is not likely to strain at such a trifle as a bit of grammar, seeing his total disregard for far more important matters—say truth, for example.

As it is not many years since Handyside kept a cheap (called in the North, "slop") boot and shoe shop in Newcastle-on-Tyne, and as he assures us through the columns of his paper that he is now a "millionaire," a little natural curiosity is excited. Perhaps, readers may think that he is going to let them into the secret of "making a pile" by quackery; he does not, however, and consequently that duty will devolve upon ourselves later on. His suggestion of how to make money is contained in a two-column notice, which he has published over and over again for some time past, headed "A Chance to Gain £1,000"; which, with the double object of insulting an honourable profession, and of depreciating other quacks' stuff, as worthless as his own, he offers to the medical profession, and the proprietors of patent medicines! In short, Handyside brags that he will pay this sum to any one of these two incongruously associated classes who can prove that Handyside's remedies are not the best yet known for the diseases which he falsely pretends to cure with them. Pending the settlement of this momentous question he says that he has deposited £1,000 in the hands of Mr. Watson, solicitor, of Newcastle-on-Tyne. We addressed a registered letter to Mr. Watson, stating that we accepted the challenge, and undertaking to show not only that the "Remedies" were not the best known, but also that they were worthless. In addition, we asked Mr. Watson to give us his assurance that the money was really in his possession, and desired him, in the event of our succeeding, to pay the said sum of £1,000 over to such local charities as we might subsequently name. But nothing came in reply for a week, and we sent another registered letter. This elicited a communication from Mr. Watson, who turned out to be the head of a highly respectable legal firm in Newcastle, informing us that he was in ignorance of Handyside's braggart challenge until he received our letter; and that no money whatever was deposited with him as stated by Handyside, so that the deserving poor of Newcastle will have to go *minus* that amount.

Handyside styles himself the "discoverer" of a Consumption Cure, a Cancer Cure, a Blood Food, and other wonderful preparations. He asserts that with their aid he has successfully treated hundreds of cases of the most virulent cancer, lupus, &c., and that "persons who have been *hacked and hewed by the profession*," (only italics will suffice him to emphasise his horror, and contemptous superiority), "so as to get at the root of this loathsome disease, have been in a short time restored to health and happiness." Referring to his untruthful challenge, he writes:—"We hope in the public interest that the medical profession, or some of its members may be found with courage enough to pick up the glove thus defiantly thrown at their feet." Instead of "glove," "empty purse" would have been nearer the mark, we thought, after receiving Mr. Watson's letter. Further on, he denounces "the sordid and selfish class whose only aim is to enrich themselves, regardless of the sufferings of their patients." How like a foul octopus this fellow Handyside is, blackening all around him, with his filthy venom, while he is searching in all directions for his prey! Yet, as to selling his nostrums, he asserts that he does not offer them to the public for the sake of profit, but "because he thinks it is incumbent on him to allow his fellow-creatures an opportunity to share in the benefits of his discovery." What a Hand(yside)some sentiment! "To attain this object he has established *consulting rooms* with *qualified assistants* (we underline these words as worthy of the attention of the General Medical Council), "in many of the principal towns, and sent vans and carriages to the less important." He says that in this way he has obtained the verdict of the people in his favour. What he has really obtained has been the money of his unfortunate dupes, mainly among the working classes in Newcastle, Hull, Leeds, Glasgow, the East End of London, and many other populous places where he has established depôts, which he styles "Temples of Health (!)" and sent his "qualified assistants." The consequences to his deluded customers may be judged from the following specimen of his illegal practice and its results.

How about other verdicts in the meantime? These he carefully ignores. Has this self-dubbed philanthropist been so absorbed in his lust of filthy lucre, that he has had no leisure for reading the newspapers; or has he been utterly indifferent to consequences so long as he could achieve his purpose? In the *Yorkshire Post* of August 30th, 1892, there is a long report of an inquest as to the cause of death of a labourer at Lockwood—Joseph Turner—who would most probably have been alive and well now, had he never heard of Handyside. This poor man was suffering from diseased hip-joint, and was recommended by Dr. Baldwin to go into the infirmary, where he would be sure of good treatment, nursing, and food; but one Richard Ayriss, Handyside's local representative, persuaded Turner's wife to fetch the invalid out of the infirmary, in order that he might be treated with Handyside's Blood Food and Liniment.

Turner rapidly became so much worse under this treatment that Dr. Baldwin was again called in, but the patient sank before he could be removed a second time to the infirmary. The doctor, in his evidence, gave it as his opinion that if the deceased had remained in the infirmary he would have recovered. The coroner, in his summing up, commented on "the folly of persons taking the advice and medicine of quacks instead of seeking the advice of regular and skilled medical practitioners," and said that the deceased's death had undoubtedly been accelerated by his taking Handyside's medicine, but that it was his own fault that he had not had proper medical treatment. The jury returned a verdict that Turner had died of blood poisoning through hip-joint disease, and that his death was accelerated by the improper treatment he had received from Ayriss.

Handyside's assistant was actually so ignorant that at the inquest he admitted that he did not even know that Turner was suffering from disease of the hip-joint. Certainly, it is scarcely to be expected that Handyside's assistants can have much medical knowledge, judging from the information given to us by a Newcastle correspondent that Handyside advertises for assistants in the *Era*, a well-known theatrical journal. Our correspondent also informs us that an applicant for the post of assistant went to Handyside, who told him that he should require him to mount a gaudy chariot, and speak to the crowd every night for two hours. The applicant suggested that such services ought to be remunerated by at least £2 a week. Handyside does not seem to have been disposed to give his interviewer a chance to "make money easy," as he would himself say, for he remarked with angry warmth, "Oh! I can get plenty of lecturers at 25s. a week, and I'll not give any more."

Handyside's own medical knowledge is, from the specimens with which he favours the readers of his

"Journal," in no sense superior to that of his assistants. It is consequently surprising how he can have become a millionaire by giving away his remedies at ridiculously low prices, as he says. But, maybe, what he loses by his liberality with pailfuls of blood food, consumption cure, &c., he makes up for by running what is called a "bucket shop" in Stock Exchange parlance; for we find in the same paper that he uses as his organ for dilating on the blessings he has conferred on mankind an announcement that "George Handyside buys and sells all railway stocks and shares that *has* no fixed value on the London Stock Exchange." N.B.—The grammar, like the boasted million, is G. H.'s own. We are not aware, by the way, of any stocks and shares that *has a fixed value* on the Exchange.

When we come to look more minutely into Handyside's writings, we are obliged to confess our inability to determine which is the more prominent trait in his character, ignorance or impudence. On reading his absolutely unfounded attacks on "a sordid and selfish profession"—one which Mr. Gladstone, in almost his last public speech, described as the most humane, most self-sacrificing of callings—impudence scores rapidly; yet, on considering his ludicrous—unfortunately, also serious—blunderings regarding the commonest medical subject, we are bound to admit that ignorance runs impudence very closely.

A theory persistently put forward by Handyside (we do not for a moment think that he believes it, ignorant as he shows himself to be), is that " creation is teeming with life-destroying microbes, *who*, from their work of *devastration* " (" Handy," as sensible people in the Tyne district contemptuously call him, is off the track again) must have teeth to destroy, and legs to perform locomotion. These microbes seize upon the lungs, and eat them away; they form cancers in the body, eruptions of the skin, lupus, and runnings of the body; they feed upon the nerves and produce nervousness; they destroy the vigour of the brain, and produce sleeplessness; indeed, nearly all diseases are produced by microbic destruction, thus weakening all parts of the body. Doctors give poisons to kill them, but these poisons weaken the whole body, and leave no vigour in the constitution to resist their wholesale invasion." What drivel! what rubbish! what falsehood!

As if Handyside had not said enough to scare his poor dupes into a stampede towards the nearest Handysidean " consulting room, van, or carriage," he gives a sketch of this fearful monster. How many thousands of poor creatures clutching at any chance of regaining lost health, how many thousands of nervous people, how many thousands of ignorant men and women have been deluded by this horrible fiction of Handyside's imagination to sacrifice their hard earnings, to sell their scanty belongings, to scatter the little hoard which they had scraped together for the time of trouble or of old age, in order to procure the nostrums which Handyside assures them will succeed where medical science is of no avail? No one

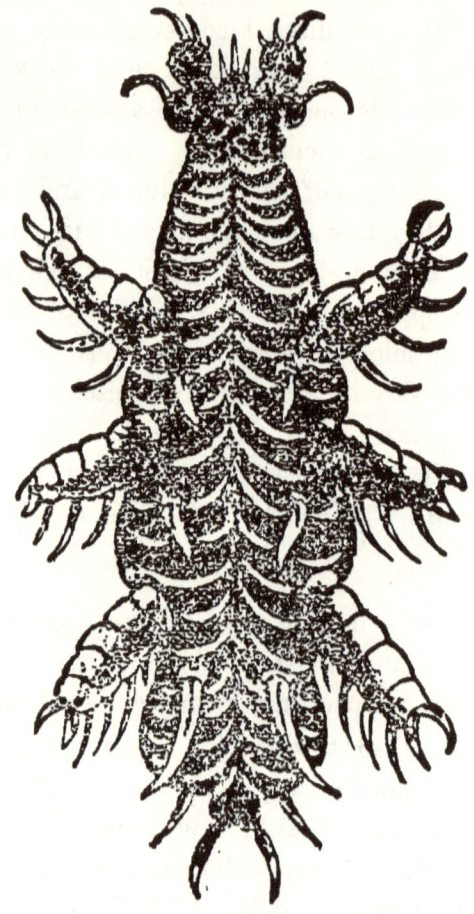

HANDYSIDE'S MICROBE (!)

can know but the Supreme Being, whose name Handyside freely uses after the blasphemous fashion of quacks, to conceal his ignorance, and to mask his designs on his dupes' pockets.

Naturally, our readers will inquire what it is that Handyside pretends to have discovered for the certain cure of the most fell diseases which afflict mankind, and that can cope with the terrific swarms of omnivorous animals which, according to Handyside, are eagerly consuming the lungs of one person, are forming cancer and lupus in another, are making a voracious meal of the nerves of a third, or running riot in the brain of a fourth. Well, if they must know, it is *Treacle*, mixed with some of the commonest drugs.

There is another Handysidean preparation which we cannot pass over, namely, the Electric Nervine Snuff. It is sold in a very tiny bottle, as befits so valuable a remedy; one which, according to the maker, " will cure tic, toothache, neuralgia, headache, pains in the head, inflammation of the gums, &c., in two or three minutes." It will also " clear the passage through the nerves " (whatever that may mean), and is " so harmless and pleasant that a child may take it." All that has to be done, says the printed directions, is "to put as much as will lie on a 6d." (making about four doses in the bottle), and fill your nostrils with it. We are progressing too fast; there is one other thing which has to be done, and that is to give Handyside, or one of his qualified assistants, 13½d. for an article which " takes the cake " for humbug and worthlessness, even amongst quack preparations, and that is saying a good deal.

We subjoin the following report by Professor Wanklyn, an analytical chemist of European reputation and eminence :—

The Laboratory,
New Malden, Surrey.

"I have made an analysis of 'The Electric Nervine Snuff' contained in the little bottle which bore the official stamp of the Inland Revenue Department.

The total contents of the bottle weighed 81 grains, and consisted of a mixture of carbonate of soda, and chloride of sodium in about equal proportions. There is a very little organic matter, which imparts the odour to the snuff."

J. ALFRED WANKLYN.

Everyone knows what carbonate of soda is; and most persons will recognise common table salt, under the technical name of chloride of sodium. The organic matter which gives a slight smell to the stuff is a trifling amount of cheap aromatic spice. Any old woman could make several pounds' weight of better stuff than Handyside's Electric Snuff out of common ingredients close at hand in her kitchen at a less cost than if she purchased one of Handyside's bottles containing only the sixth part of an ounce.

The action of the moist mucous membrane lining the nostril, when a pinch of this precious compound —precious only, at Handyside's exorbitant charge—is snuffed into the nose, is to cause the salt to melt slowly, and thus create a temporary sensation of coolness. Its continued use would seriously affect the sense of smell, and produce other injurious effects.

There is no more electricity in the stuff than there is modesty or honesty in its "discoverer." It suits Handyside to call it Electric Nervine Snuff. We should designate it by a much shorter and more correct name. We have often heard the remark made that the Patent Medicine Stamp is practically a license to swindle. It would be quite unnecessary to go further than Handyside's Electric Nervine Snuff to fully prove the truthfulness of this observation.

Chapter XII.

Mattei's Electricities in Court; a Curious Way of Exposing Quackery.

"*Oh, dear! What can the Mattei be?*"—
New Version of an Old Song.

In one of the earliest of the series of articles, now reprinted*, we found it our duty to comment strongly on the so-called "Electricities," bearing the name of "Mattei," and we showed, by the analysis conducted by Mr. A. W. Stokes, public analyst for several important metropolitan districts, that all of these, though fantastically distinguished by the vendors as the blue electricity, the red, and the green (this last possibly, out of delicate compliment to the Matteist dupes, who believe that a few drops of these precious remedies will suffice to cure cataract, or to unite broken bones), consisted solely of water. Mr. Stokes' analysis was thoroughly confirmed by an investigation made quite independently by Professor Michaud, principal chemist in the Cantonal Laboratory of the Department of Justice at Geneva, whose report was indentical with that supplied by the English analyst.

See Exposures of Quackery, Vol I., page 15.

Water, "nothing more!" Precious, as the Matteists call it, without doubt, seeing that as much is charged for a small phial containing an ounce or so as one would have to pay to a water company for 5,000 gallons.

The revelations we published failed to give unalloyed gratification to Mattei's London agent. Some of our readers might think patent medicine men would be the last people to object to a widespread gratuitous advertisement, so partial are they to advertising their nostrums anywhere and everywhere—in newspapers, pamphlets, and handbills, on walls, omnibuses, bathing machines, and even the sails of fishing-boats, alongside the railways, down low on the pavements we walk over, or high up amongst the clouds, by means of electric flashes. But there is a time when they distrust, and even detest, publicity, and that is apparently when Truth comes on the scene. Otherwise, we are at a loss to understand why we have been the recipients of so much abuse (generally anonymous, yet referring to some particular article), and of so many threats of legal consequences after exposing the sham pretensions of this or that patent medicine.

To return to Mattei's London agent, however, He sent us the usual deprecatory letter, concluding with threats and hints of dire results should we venture again to say a word in disparagement of the Mattei preparations. Our best and only answer was to give his letter the same publicity in our journal that had been given to the analysis and reports upon the worthless Mattei "Electricities"; and we used the same heading for our reply that appears at the commencement of this editoral note. It is almost needless to say that we received no more communications from Mattei's London agent, or from any legal quarter, although the following mysterious announcement appeared in an evening paper:—"Count Mattei's solicitors, Messrs. Colyer and Colyer, say that the Count has determined to defend his medical system against the attacks which have lately been made upon it by numerous English journals, and that a writ for an action for libel has already been served upon one of them." We can only observe, with regard to this terrible threat, that either Mattei's black list is a very long one, so that our name has not yet been reached, or that the single writ alleged to have been issued must have been withdrawn, for no such action as is grandly announced has been tried in the High Court of Justice, although two years have elapsed since the notice appeared.

Mattei "Electricities" have cropped up lately, —broken out in a new place, as it were,—for we read, in a Derbyshire newspaper, which one of our subscribers kindly sent to our office, a curiously significant report of some proceedings at the Worksop County Court. A woman named Blakeley, styling herself a "female physician," sued a number of people whom she had supplied with "the Mattei remedy," but several of the defendants asserted that the pretended remedy was absolutely worthless, and in support of their statements handed to the judge sundry bottles containing samples of the fluid, with which the "female physician" had supplied them. His Honour, Judge Masterman, adjourned the cases in order that the county analyst might examine the samples and report to the Court at the next sitting, Before that time, however, the plantiff had withdrawn the actions. "Women were deceivers ever," as the old song says, and the "female physician" may have displayed only the proverbial fickleness of her sex; but it is a very suggestive coincidence that the county analyst reported that he had not discovered any other substance than water in the bottles.

Water, "honest water," as Shakespeare called you in *Timon of Athens*, to what base uses may you be put when you fall into the hands of a quack!

We have received letters from many correspondents directing our attention to the fact that a certain weekly paper has lately assumed to itself great credit for its exposures of quackery; following apparently in the wake of articles, some of which were published by us years ago, indeed, even previous to the existence of the journal in question. We are glad to find that another journal is taking the matter up; though, perhaps, we might feel still more pleased if its editor would do us the ordinary courtesy to acknowledge his obligations to us.

Nor can we say that we appreciate his method of treating quack preparations, seeing the unblushing announcements with which the advertisement columns of our contemporary, *Science Siftings*, abound. One of our correspondents states that he counted a dozen such in one number. Quite likely, for in the latest issue, containing an adverse article concerning Clarke's Blood Mixture, we came across advertisements of Eno, Beecham, Holloway, and other notorious patent medicine men. Curiously enough we observed no advertisement of Clarke's Blood Mixture. Is this to be regarded as an instance of Cause and Effect?

Presuming that *Science Siftings* is in earnest in exposing quackery, the appearance of Eno's, Holloway's, and Beecham's advertisements must have arisen through some inexplicable accident. However this may be, we are sure of one thing, namely, that we shall receive fewer communications from correspondents somewhat incredulous as to its sincereity, if, having followed in our track so far, *Science Siftings* will go a little farther, by adopting our rule, "Quack advertisements rigorously excluded," and acting up to it. Its statements would then be entitled to more respect, and would carry more weight with people who now comment—rightly or wrongly, we do not profess to decide—upon the singular anomaly of a journal attacking a few quacks and tacitly aiding and abetting a crowd of others, by inserting their advertisements.

Chapter XIII.

PATENT MEDICINE TESTIMONIALS; ST. JACOB'S OIL; CLARKE'S BLOOD MIXTURE; THE MAN WHO GAVE HIMSELF A TESTIMONIAL; ENO'S FRUIT SALT.

"*Men often swallow falsities for truth.*"-
SIR THOMAS BROWNE, M.D.

THE worthy old Norwich doctor, whose philosophic writings are as sound now as when they were penned more than 200 years ago, would have found little difficulty in proving this axiom; for he would need only to direct the attention of any sceptical individual to the first batch of quack puffs and testimonials that might come to hand. We have already devoted more than one chapter in "Exposures of Quackery" to this subject; but scores of pages might be written upon it, if we cared to pursue such a theme. In dealing with quack medicines generally, we feel somewhat in the position of a guide conducting a stranger across a bog; we need only indicate the rottenest parts, and leave it to the good sense and intelligence of our companion to form an accurate opinion of the rest.

Quack testimonials may be divided into three classes, those given in error, the probably false, and the palpably forged. The persons who furnish the first class are almost invariably of the most ignorant section of the community. Their reasoning power is as feeble as their grammar; yet they assume an off-hand knowledge of the nature of diseases which would be almost presumptuous in a physician or surgeon of great experience. They are as ready in diagnosis as if all their lifetime had been given to special study, and the result is that an ordinary cough is magnified into consumption in its last stage, that an obscure tumour is declared to be cancer, and that any common indisposition is, in their estimation, a serious illness. An advertisement happens to attract their notice; they swallow the contents of a box of Fleccem's pills, or they diligently rub in a potful of Gulloway's ointment, and, the cough disappearing, or the swelling subsiding, they communicate the wonderful fact to some patent medicine man, who spends hundreds of pounds in publishing to the world that A.B., or C.D., has been cured of consumption, cancer, &c. The poor ignoramus who gives the certificate, has made a double blunder, in mistaking the character of his ailment, and in overlooking the fact that any aperient or brisk rubbing, as the case may be, would have soon set him right.

It is very unfair, to say the least of it, that the patent medicine man carefully touches up the spelling and grammar of every such testimonial before putting it into print; since the public are deprived of an opportunity of judging what manner of individual it is who has given the certificate. Some time ago, we read such a remarkable account of a "miracle" claimed to have been wrought by St. Jacob's Oil (turpentine and camphor, scented) that we thought it incumbent on us to obtain some supplementary information. The narrative, as detailed by a grateful father, was to the effect that a young man, living in Leicestershire, had been suffering for five years from a mysterious rheumatic affection, that he had been under the treatment of many doctors, as well as at various hospitals, and that "the most celebrated medical men of the Midland Counties" (a large order, this!) had pronounced his case to be incurable. A few bottles of St. Jacob's Oil were bought on the recommendation of a sympathising neighbour—our readers will begin to guess the rest—and in a short time the young man made a complete recovery. We wrote to the father, of whom we expected an intelligible reply, for we had not then dipped so deeply into the patent medicine well, in our quest of Truth. What we did get was the following communication:—

"In ancey to yours of this morning which we recevd, you ask me to state the case of my son, the tretment he recevd at Leicester was at the infermery and from an old Docker which has past away since then, and then he" (the son we imagine, not the Docker), "was sent to the Devenshire Hospital, Buxton, and from their he was sent home quite a cripple on cruthes, and was al formes and then i gat the oil and i have the same greate faith in them as i always have done."

We were so utterly prostrated by our attempts to make out what all this meant, that we were "al formes" before we came to a conclusion; and fearing that further correspondents might bring us to the same desperate pass as the other "Docker," we deemed it prudent to discontinue it; besides which, it would have been obviously an unequal contest, to put the testimony of any number of "dockers," hospital or private, against that of a person of the deponent's learning and scientific attainments.

We need not dwell at any length upon the probably false class of testimonials. They have a strong family resemblance; they seldom descend to such trivialities as names, addresses, and dates; but they make up for these deficiencies by constant assertions of many years' continuous suffering on the part of the patient, and of similarly long period of impotent ignorance on the part of medical men who were so profoundly stupid as not to know that Shegull's Treacle or some other quack nostrum would have cured the case in a jiffey.

But it is when we come to the palpably forged certificate that we begin to realise some idea of the extent to which a patent medicine man will go. Until some two or three years ago the lengthy advertisements of Clarke's Blood Mixture wound up with the following testimonial :—

"*Clarke's Blood Mixture is entirely free from any poison or metallic impregnation, does not contain any injurious ingredient, and is a good, safe, and useful medicine.—Alfred Swaine Taylor, M.D., F.R.S., Lecturer on Medical Jurisprudence and Toxicology.*"

We were astonished on reading this; and we were certain that not a single word of this certificate was ever penned by the late Dr. Swaine Taylor. As a matter of fact he had, during his lifetime, exposed the pretensions of Clarke's Blood Mixture in a letter published in the *Lancet*, 1875, in which occurs the remark that, "The sale of medicines of this kind should be strictly prohibited, unless the bottles containing them were issued with a caution label setting forth their true composition." It is only reasonable that a person should know what he is purchasing." We invited the proprietors of Clarke's Blood Mixture to answer four plain questions, easy enough for any straightforward person to deal with, viz. :—1.—When, where, and under what circumstances, did Dr. Swaine Taylor give the alleged testimonial ? 2.—By whom was the signature witnessed ? 3.—When and where could the original be inspected ? 4.—Why did the proprietors of the Blood Mixture withhold from the public knowledge so important a document until years after Dr. Taylor's death ? Though we registered this communication, as we did several subsequent letters, they did not venture on a reply. They suddenly withdrew the bogus certificate, which, for many months, they had paraded in every newspaper throughout the land ; but not a word of answer or explanation did they vouchsafe. They had none to give. The whole affair was palpably false; their conduct and their silence under exposure, showed it, and, later on, we incontestably proved that the testimonial was a forgery. The story is too long for the present chapter. but it has appeared in other articles, and will be found in detail in "Exposures of Quackery." The proprietors of the self-dubbed "world-famed" Blood Mixture, boast of their thousands of "wonderful" testimonials. What is there "wonderful" in a falsehood, even if it be repeated thousands of times? The only thing wonderful about this particular one is that its perpetrators have been additionally guilty of a cruel, revengeful outrage upon a dead man's spotless reputation; and this alleged certificate which we have demonstrated to be a deliberate untruth, makes one very sceptical as to the origin of the others. In France, Germany, and some other continental countries, the authorities would, on the circumstances being made known, have taken steps to punish the perpetrators of such a public fraud*.

From such a repulsive subject we turn with some relief to what may by termed the humorous aspect

---

* "An excellent paper, the HEALTH NEWS, has been publishing analyses of most of the patent medicines now being put before the public, and the result is not a little startling. Not only are these nostrums actually useless in every case, from a medicinal point of view, but in the majority of instances they are actually injurious to the brainless idiots who purchase these pills, lotions, and mixtures. Further, in some cases bogus testimonials are used to bolster up these worthless compounds, and other scandalous means used to deceive the public in the matter. In one or two cases the services of the Public Prosecutor seem urgently needed."—*Whitehall Review.*

of patent medicine testimonials. Last year, when typhoid and other fevers were prevalent, there appeared in some of the London daily papers a curiously worded paragraph, inserted amongst the reading matter, but evidently to the initiated eye a flimsily veiled advertisement, headed "The Dowry of a Nation." It began as follows :—"The names and memories of great men are the dowry of a nation in death as well as in life." A noble sentiment, truly, though rather mixed; for the *memories* (using the word in the sense implied here) of any men, great or otherwise, could not exist during their lives. But we read on, hoping to be rewarded by an anecdote of some great man. We were disappointed, for the only individual mentioned was one who would hardly be accepted by any nation as the equivalent of a dowry, whether living or dead; unless he should imitate the late Holloway, and endow a lunatic asylum as a fitting memorial of the implicit faith which his credulous customers had in the remedial powers of his patent medicine, or, rather, in the statements made in his advertisements.

"I used my Fruit Salt freely in my last attack of fever, and I have every reason to say it saved my life. —J. C. Eno." Here, then, is the dowry of a nation; here, then, is the great man so delicately alluded to in the opening sentence. There would be as much difference, comparatively speaking, between buying J. C. Eno at our valuation and selling him at his own, as there would be between buying Epsom Salts at their market cost, and selling them at the price charged for Eno's Fruit Salt; except that in the latter case the transaction would be nearly all profit. But J. C. Eno has proved himself great in one respect; he has taken the wind out of his envying rivals' sails, and beaten the record by giving himself a testimonial. Unfortunately for medical science, of which J. C. Eno poses as a shining light, the particulars furnished of his interesting case are too mysterious and brief to be instructive. We can understand his using his own nostrum freely, though it is very costly to other people, considering the commonness and cheapness of its ingredients; but, to enable any one to profit by his experience, he ought to have stated—if he knew—what kind of fever it was that he was suffering from, and what was his reason for thinking that his stuff saved so valuable a life. He speaks of the illness as his last attack, from which we are led to conclude that he has had previous illnesses from the same cause. Why J. C. Eno, with such a miracle-working preparation at his command, should have allowed himself to have more than a single attack, why he should not have cut short such an objectionable tendency to fever—unless he had his misgivings that his treatment might cut short J. C. Eno, too—we are equally at a loss to understand.

Still, we cannot help admiring his enthusiasm, or his rashness. Judging from our experience of forty years' medical practice, we should hesitate to administer, either to a patient or to ourselves, if suffering from typhoid fever, say, large doses of Epsom salts; in the former case it might lead to a verdict of Manslaughter from a coroner's jury, in the latter to a verdict of "Serve him right," from professional men, who presumably know something about such matters. J. C. Eno is not a doctor; he belongs to a class that finds easy advertising and indiscriminate drugging more remunerative than hard work and discriminative diagnosis. "Where ignorance is bliss, 'tis folly to be wise," a quack may think, when Judges, in summing up, act upon the principle that "Ignorance excuses" manslaughter by quacks. Yet, "*Ignorantia non excusat*," is a favourite maxim when a man pleads, in extenuation of his offence, that he was ignorant of Law.

Eno's much advertised Fruit Salt consists of the following inexpensive ingredients:—Carbonate of soda, tartaric acid, Epsom salts, bicarbonate of potash, and sugar. When a tablespoonful of this powder is put into water and stirred up, effervescence is caused. The only constituents which are in the remotest degree connected with fruit are the tartaric acid and sugar; but the sugar is that obtained from the sugar-cane and not from fruit, while the tartaric acid, carbonate of soda, bicarbonate of potash, and Epsom salts (sulphate of magnesia) are readily procurable in enormous quantities, at trifling cost, from the waste products of various chemical manufactures. There are few persons who have not some knowledge of these articles in their undisguised condition, and fewer still who would be so idiotic as to imagine that Epsom salts grew on gooseberry bushes or at the top of cherry trees.

The title of "Fruit Salt" is entirely a misnomer, and as deliberately misleading as Eno's untrue assertions that such simple substances as we have just mentioned, will cure every conceivable complaint, —the "great man's fever" included,—or as his gaily coloured pictorial advertisements, in which pines, melons, peaches, and other choice fruits are profusely displayed, in order to make people believe that these horticultural and greenhouse treasures are utilised in making the "Fruit Salt." No fruit whatever enters into that stuff; and none, indeed, ever enters into Eno's Hatcham Works save, perhaps, now and then an orange in the pocket of some small boy engaged there in the lucrative process of converting Epsom salts into "Fruit Salt." "Fruit" is a short, catchy, and catch-penny word, but the Chinese have a shorter, and more expressive term when they call a certain spurious article, "Lie" tea!

Chapter XIV.

"Handy" Still Shuffling.

"*A fool is nauseous, but a coward is worse!*"—Dryden.

"*Striving to make an ugly deed look fair.*"—Shakespeare.

When it came to our knowledge that Handyside, of Newcastle, had put forth in his obscure puffing paper, his challenge (!) to the medical profession, we took it up, as already described in chapter XI; but we soon found that it was an impudent piece of brag, and that he had falsely stated that the sum of £1,000 was staked in good keeping. A week after the publication of our article, Handy summed up sufficient pluck to write to us, goaded thereto, we imagine, by the vigorous remonstrances of a highly respectable firm of solicitors at the unwarrantable employment of their name. They should have prosecuted him for using it fraudulently. Of course, after our serious charges any one would have supposed that the delay might be due to indignation too great for utterance, and that "Handy's" letter would be couched in angry terms. Not so, however, but a tame epistle, the only bad language in which were the shockingly bad grammar, and the equally defective spelling characteristic of Handysidean literature. "Handy," with all his bounce, knew his book too well to invite additional exposures of his worthless stuff by contradiction, and he merely talked round the question, without touching it. The "wrathful dove," or "most magnanimous mouse" would have been more energetic in the defensive.

Here is "Handy's" letter :—

"12, Bentinck Crescent,
Newcastle-on-Tyne,
Oct. 6th, 1894.

To the Editor of health news.

Dear Sir,—You have wrote to Messrs. Watson of this City concerning a Challenge I made. I deposited with them check for £1,000 and challenged all the Doctors in the world when the challenge was not accepted I withdrew the check but I am ready to deposit it again if you will deposit the same and have it tested in all the principill towns in the Kingdom with all the doctors in the world to assist you. This is not a medicine it is simply unrefined boiled vegetables sweetened up to make it as palatable as possible.

Yours truly,
G. HANDYSIDE.

It is a pity that "Handy," who is as unrefined, apparently, as the boiled vegetables which have raised him from the "slop" boot and shoe shop of his earlier days to the position of the proud millionaire (as he designates himself), does not keep one of the "qualified assistants," whom he locates at his numerous consulting rooms, at Bentinck Crescent, to write his letters for him, or, failing that, bribe a boy from the nearest

Board School with an occasional sixpence to act in that capacity. Evidently, he is too much occupied in writing his *checks*, and in his bucket-shop transactions, to begin learning common grammar and spelling now. "Principill" is a humorous rendering of the word, as coming from a quack; though not so excruciatingly funny as "Handy's" invitation to all the doctors in the world to take part in a tour—personally conducted by "Handy," we presume—through all the "principill" towns in the kingdom, for the purpose of assisting us in testing his "unrefined boiled vegetables," *plus* unrefined treacle. The test has been already accomplished, and we do not desire any of "Handy's" transparent equivocations. Even his boy-secretary could tell us that treacle was obtained by boiling vegetable matter, in the shape of sugar canes.

But, on trying to extract some little sense from "Handy's" epistle, we find that he has in his ignorance set us an impossible task, viz., to deposit the *same* cheque that he deposits, and then to have this same cheque tested by all the doctors in the world. It would be easier to expose all the unrefined quacks than to accomplish the impracticable feat suggested by "Handy." And, after all, what does "Handy's" letter, when done into English, amount to? To a deliberate ignoring of the terms of his original challenge, to a miserable shuffle and equivocation, and to a transparent effort to hide his discomfiture behind empty brag. If he means business, as the sporting papers say, if he is really honest in his offer, let him place, not a *check*, but Bank of England notes, in responsible hands for the specific purpose of paying it over to our nominees when we have proved both that his vaunted remedies are not only not the best, but that they are worthless! It would not be long before the £1,000 were distributed amongst deserving Newcastle charities. The money would, having regard to the unscrupulous way in which he has made it, be filthy lucre, indeed, when it left his possession "for aye and for ever," but the dross would be (to borrow his own term) refined by the use to which it was applied. Further, have we not a precedent for accepting it, regardless of the manner in which it had been acquired? Certainly; for on a particular occasion, when General Booth was criticised for receiving a donation to the Salvation Army funds from a peer of very questionable character, the General retorted with considerable emphasis, that if the Devil himself sent a contribution, he (the General) should feel justified in taking it, on account of the noble use to which it would be put.

Like a whipped cur, doing a stealthy growl at a safe distance, "Handy" has since issued a handbill, headed "The £1,000 Challenge likely to be accepted by a Doctor and Editor." We make it a rule never to withhold praise where due, and we therefore hasten to congratulate "Handy" on having apparently followed the advice we gave him in HEALTH NEWS, and having engaged a boy from the nearest Board School to assist him in his literary compositions. "Oh, dear, what a surprise!" we thought, as we rubbed our puzzled eyes. "*Check*" is now spelt "cheque;" "*principill*" has been translated into "principal," and other marked signs of improvement are present. But though "Handy" does not wield the pen, he spoils the grammar by dictating the words; and "them's my sentiments" is the all-pervading note of the illiterate and incoherent document. Here and there "Handy's" ideas are as thick and dark as the common black treacle, which constitutes the chief ingredient of his "cure"—to such an extent, indeed, that a Newcastle correspondent tells us that he has seen as many as nine barrels of it delivered at "Handy's" place at one time. On another occasion, when a treacly consignment arrived, a bystander sarcastically asked "Handy" (prophets have no praise in their own country, especially those of the Handysidean stamp, who would doubtless spell the word "profits") why he did not use black beer, as a change from the perpetual black treacle? "It would ferment too much," responded the artful "Handy," and, he might have added, it would cost too much.

To return, however, to "Handy's" manifesto. He still shuffles and prevaricates over the terms of the challenge; as formerly printed they were plain and short, now there is a long rigmarole, as follows:—"The terms of the contest are as follows: That all the principal towns in the kingdom be visited once a fortnight, and Handyside's Consumption Cure be given to those who require it, and doctors' medicine to be given also to all who need it. At the end of twelve months all the patients to assemble. Those who have taken Handyside's Consumption Cure to stand on one side, and those who have taken doctors' medicine to stand on the other. Each patient to make a statement of what each medicine has done for them, and let the majorisy" (we think "Handy means "majority," but we must give the terms as they are printed) "decide. This surely must be fair and honest, for the people must be the best judges of what has done them most good."

This rigmarole is merely a sort of dust-throwing in the eyes of his poor deluded dupes, who are beginning to realise the extent to which they have been humbugged by "Handy," and it would not take any close study of the *new* terms to show that it "surely must be fair and honest" on "Handy's" part, if he had any faith whatever in the healing properties of the contents of his treacle-tubs, to stand by his original challenge.

We have already stated that his fresh proposition that "all the doctors in the world" should visit all the "principill" towns in the kingdom," to assist us in testing his stuff was ludicrously impossible. Stay! perhaps "Handy" may be as ignorant of the number of principal (*i.e.*, large) towns in Great Britain and Ireland as he is of many other matters on which he might with advantage consult the Board School boy.

We have a fair amount of energy and endurance; but whether travelling with all the doctors in the world for twelve months—hey! how railway shares would jump up in prospect of such an extra amount of traffic—or alone, we could not, nor could "Handy" himself, journey to all the important towns in the kingdom once a fortnight, still less twenty-six times in the course of the year. Besides this drawback, there are other points on which "Handy" has evidently not taken the boy into his confidence, otherwise, that intelligent youngster would have suggested that, since the exposure in HEALTH NEWS, there would, if lunatic asylums were barred from participating in the contest, be no small difficulty in finding enough people to "require" "Handy's" stuff; common black treacle and water, made to taste more physicky by flavouring the mess with a common bitter would not charm the multitude, who are beginning to get a little wiser already, judging by the fact that when we recently passed one of "Handy's" Temples of Health—the misleading name he gives to his shops—about eight o'clock p.m., on a "free distribution night" (locality, the Commercial Road, E.), there was only one patient visible, a scared-looking youth, apparently contemplating a bolt as soon as the "qualified assistant" (Handy's term for a shopman) had turned away to reach a sample bottle from the shelf.

Again, a still greater difficulty than those we have referred to presents itself. If, knowing what we do of the composition of "Handy's Cure," and its uselessness in the treatment of cases of actual cancer and consumption, we were to countenance its administration to any person suffering from either of these diseases, and that person died, we should most assuredly lay ourselves open to the criminal charge of being accessory to manslaughter, while "Handy," not being a qualified medical practitioner, would have the benefit of the doubt, because, being ignorant of medical matters, it would be assumed that he had acted to the best of his ability. Such is the glorious state of English Law at the end of this enlightened nineteenth century.

"Handy's" dodges and mean attempts to shuffle out of the consequences of his original challenge, and thus to deprive the Newcastle poor of some restitution of what he has robbed them of will not wash. He speaks glibly of what is "fair and honest," yet all the while he must be conscious that he is talking utter nonsense, a sort of goose-step of rhetoric, to make belief that he is coming to the point. There are two ways in which the matter can be settled, and which will commend themselves to every sensible person. One is to have his stuff tested by a previously agreed number of skilled analysts, and to abide by their decision. The other, should he prefer a legal to a scientific verdict, is to proceed against us in a court of law, and let a jury determine which of us is in the right. But he must previously deposit the £1,000 in some responsible individual's hands, in the form of bank-notes or coin, not "a *check*," as he offers to do. We should not care to run the risk of having to waste our time over a second action about his cheque.

As regards the rest of his windy and evasive answer to our exposures, we have only space enough left to state that it is all ridiculous, with the exception of those portions of it which are false. We notice that he carefully avoids giving the name of our journal, presumably because he is not particularly anxious to make it known, lest people should read our articles for themselves, aud not be content with his garbled and incorrect version of them.

CHAPTER XV.

QUACK NEWSPAPERS; THE SEQUAH BUBBLE BURST;
TRANSLATION OF "EXPOSURES OF QUACKERY"
FOR INDIA.

SEVERAL of our readers have forwarded to us specimens of bogus newspapers circulated by patent medicine men for the purpose of puffing their wares; all of the same deceptive character as those which we described in a recent article in HEALTH NEWS. The last of these which has come to hand is called the *Evening Telegram*. No printer's or publisher's name is given, and the only address and date vouchsafed are "London, October, 1895." It is styled "No. 6, Vol. V.", and the price marked on it is 2d. Why this sum more than any other, it is difficult to say. Possibly, the editing was left to an office-boy, and as no previous issue of the *Evening Telegram* had ever been published, notwithstanding the deliberate falsehood contained in "No. 6, Vol. V.", the boy, recognising the fact that the broad-sheet was worth nothing, repeated to himself the vulgar saying, "Twopence more, and up goes the donkey!" and putting 2d. more than its value on the front page, left the rest of the matter to the donkey, namely, that section of the British Public which is

always ready to swallow any quantity of lies and of
quack nostrums. How many millions of No. 6, Vol. V.
Hood's Sarsaparilla Syndicate have dosed the poor
donkey with we cannot say; but we may mention that
on the same day that our subscriber sent us a copy, one
was thrust into our letter-box, and we, later on, saw
men actively distributing gratis this curious evening
newspaper, the only No. that the publisher could produce,
even if he were offered £1,000 instead of 2d. for any
other published previous to it. It is profusely illustrated with portraits of people who are said to have
experienced wonderful results. "Hood's Own," as
they may be classified, describe their cases in the usual
quack literature fashion. One woman was cured of
"hardly daring to go upstairs;" another announces
that she is now equal--by her portrait we should not
imagine she is very willing—to doing house cleaning;
while a third assures us that she, too, is now able to
work. From her likeness, which appears to be a copy
of that of Her Royal Highness the Princess of Wales,
we should not think that she has any occasion to put
her powers to the test. It is something, though, to
know that this delicate creature, who tells us that she
was formerly a martyr to "nervous spells," and that
her stomach "bloated" after eating, must be a thorough
woman of business, for she winds up a long letter, not
with a feminine P.S., but with an emphatic "N.B.—
Take only Hood's." Anyone would be disposed to
judge from the peculiar finish to this letter that Miss

Smith, of America, had a direct interest in pushing the pills and mixture. Interspersed with the gushing testimonials in the *Evening Telegram*, all presenting such a strong family resemblance that they might have been written by the same individual, there are some articles of a general character. The longest of these is headed, " In Scotland Yard," and that locality is such an appropriate one for a sham newspaper publisher to find himself in, that we will leave him there.

We turn with a feeling of relief to another paper, the *Financial Times* of October 3rd, which has an amusing and instructive leader, entitled " The Sequel to Sequah." From this we learn that Sequah, Limited, has contrived to get rid of a quarter of a million of pounds sterling extracted five years ago from the gullible British Public. In return for this magnificent sum the shareholders obtained the proprietorship of the Sequah Prairie Flower Mixture and Sequah Oil; both of which, as we have proved in " Exposures of Quackery," Vol. I., are utter rubbish. Our first article on the Sequah stuff was published in 1892, and some little time afterwards, at a meeting of the Company, the chairman gave us a gratuitous testimonal, as to the public service we were doing, for he attributed the falling position of the Company, and the absence of any profits (in 1890 it paid 14 per cent.), to the injury which we had done it. Gross ingratitude, this, in return for the widely spread advertisement which we made the Company a present of! But, then, there are times when even a patent medicine man shrinks from publicity, and this was one. We gave the analyses by Mr. Stokes, F.C.S., of their common aloetic compound ("Prairie Flower") and of their "fishy" oil; we showed that their pretended Indian lecturers were only ignorant, impudent "cheap Jacks;" and the company fell, like Humpty Dumpty, beyond possibility of reconstruction. It has now reached the final stage of liquidation.

Speaking of the Sequah sham North American Indians, we are reminded of a letter, dated Sept. 17th, 1895, which we have received from a large firm in Lahore, India, Sada Nand, Brothers, asking permission to translate various portions of our published "Exposures of Quackery" into the vernacular tongue, with a view to circulating the facts therein contained, very widely amongst the native population, so as to open their eyes to the real character of the English and American quack medicines which are now sold enormously in India. We have given the desired leave with much pleasure, and we cannot help, at the same time, contrasting Messrs. Sada Nand's courteous request with the conduct of some London papers, which have appropriated our articles for some years past without the slightest acknowledgement, or taking any further trouble than to flimsily disguise the matter before publication.

Chapter XVI.

Fenning's Fever Curer.

"All maladies . . . all feverous kinds."—Milton.

These few words, quoted from the great poet, convey a concise idea of the many ills to which poor human nature is subject. But great men are neither limited to poets nor to any particular epoch, so that it happens that, at the present day, there exists another great man who, according to his own assertions, can get rid of "all maladies. . . . . all feverous kinds" in an incredibly short space of time. This remarkable individual does not seem to be in the least degree troubled with the modesty usually found in connection with real genius, for he puffs his wares extensively, and will, therefore, doubtless be obliged to us for giving "bold advertisement" (gratuitous, too) to his name—Alfred Fennings, West Cowes, Isle of Wight. Our enquiries into his medical credentials have, as we expected, led only to negative results.

However, let us see what he has done, and what he can do—only according to his own account, be it understood. We "dare do all that does become a man," but we must ask our readers not to make such a great demand upon our courage—or our credulity—as to require us to vouch for a single iota of Alfred Fennings' assertions.

Well, he is the author of a pamphlet called *Everybody's Doctor; or, When Ill, How to Get Well*, a glance at the contents of which suffices to show that all that is requisite is to take Fennings' This or Fennings' That. As to what he can do—of course, we mean what he says he can do—here is a specimen. He is describing the marvellous effects of one of his nostrums, Fennings' Fever Curer, and he tabulates them as follows:—

> "Bowel complaints cured with one dose.
> Typhus or low fever cured with two doses.
> Diphtheria cured with three doses.
> Scarlet fever cured with four doses.
> Cholera cured with five doses.
> Influenza cured with six doses."

This somewhat rhythmical arrangement of the boastful account of Fennings' triumphs has set our pen quite on the jingle, and we must consequently crave our reader' indulgence while we put his achievements into verse:—

(Air: "Ten Little Nigger Boys.")

Six little Cowes boys,
   Influenza seized one;
Fennings gave him doses six,
   Influenza quite had gone.

Five little Cowes boys,
  One the cholera took;
Fennings gave him doses five,
  And brought the disease to book.

Four little Cowes boys.
  One had scarlet fever dire;
Fennings gave him doses four,
  And thus subdued the fire.

Three little Cowes boys,
  One diphtheria caught;
Fennings gave him doses three,
  Diphtheria went to naught.

Two little Cowes boys,
  With typhus one laid low;
Fennings gave him doses two,
  Away did typhus go.

One little Cowes boy,
  His " stummy " felt so bad;
Fennings gave him but one dose,
  And that settled the —

Confound it! Our pen has suddenly become prosaic again; neither " stomach-ache" nor " bowel complaint" will rhyme to " bad," and we have much further to journey before we have completed our task—before, indeed, we have enumerated half of the virtues of the Fever Curer, according to its inventor and maker. For after giving a lengthy list, printed in capitals, of diseases, for which this " celebrated remedy " is a sure cure, he says :—" Fennings' Fever Curer will *prevent* persons catching typhus or low fever, scarlet fever, cholera, diphtheria, and yellow fever, or will cure with two or three doses all these fearful diseases. No case of typhus fever is *hopeless* with this remedy at hand." Should any one be so sceptical as to demand proof, the only reply to such an utterly lost person is that given by the maker himself—in italics, too: *See Fennings' Everybody's Doctor, page 8.*"

A certain cure for all these maladies is what one may term a large order; but a certain preventive is still more upon the wholesale scale. We have not a copy of *Fennings' Everybody's Doctor*, so that we cannot ascertain where Fennings professes to have gained his assumed knowledge of yellow fever, a disease which is fortunately unknown in these islands. In a long professional retrospect, we can only call to mind one outbreak in this country. That occurred at Swansea in September, 1865. A vessel named the *Hecla* arrived in that month from Cuba with one of her sailors dying, and two others convalescing, from yellow fever. Between September 15th, six days after her arrival, and October 4th, twenty people in Swansea having business relations with the *Hecla* were attacked with yellow fever, as were also three more sailors on board of a small vessel which had been lying alongside that ship. Fennings may have seen yellow fever abroad, as we did many years ago, in the Southern States of America; but, if he ever did see a case of that fearful disease, he is something far worse than mercenary in trying, for the sake of filthy lucre, to delude people into the notion that his stuff can either prevent or cure yellow fever.

We remember on one occasion having charge of a number of patients stricken down with it; one was an old Mississippi boatman, whose experience of the malady was greater than ours (or even Fennings') ever will be. With the desire of giving comfort where cure was evidently out of the question, so fatal was that particular epidemic, we tried to soothe the doomed man's last hours with a few words of encouragement. Starting up in bed and turning his face full towards us, he shouted almost savagely, "Yer ain't going to bluff me like that!" then, falling back exhausted by the effort, he added, in a soft undertone, almost in a whisper, "Guess, Younker, yer ain't seen any case of yeller fever before now?" When we went to the improvised tent hospital early on the next morning, the poor fellow was dead and buried, for the fatality was as swift as it was certain.

Does Fennings mean to insist upon the assertion that his stuff can cure yellow fever, typhus fever, cholera, small-pox, and other affections? If so, we can only say that he is not going to bluff us like that. But, if he persists in the further assertion that his Fever Curer will prevent people catching these diseases, then we must designate that statement as the grossest exaggeration, or what many would call by a much shorter name. In his gamut of the curative properties of the Fever Curer, he carries up to what six doses will do.

If seven would only cure lying, what a boon it would be in his case and that of other quacks; some of whom do not even stick at putting forward bogus testimonials with dead men's names attached, to make them look genuine. (See Chapter III.)

But it is no use to sigh for the moon, or to expect even ordinary beneficial results from the stuff described in the following analytical report sent to us by Professor Wanklyn :—

"The Laboratory, New Malden, Surrey.
"Fennings' Fever Curer is a dilute solution of nitric acid (strength $1\frac{3}{4}$ per cent.). It is flavoured with peppermint, and contains *minute traces* of other organic matter, including a very small amount of opium.
J. ALFRED WANKLYN."

Upon the wrapper in which the bottle that we submitted to Professor Wanklyn for analysis was printed, we observed, the following notice :—"An Act of Parliament requires that the word 'Poison' should be printed on each bottle of Fever Curer, because it contains a few drops of laudanum, but which becomes, by being mixed with certain liquids, a most beneficial physic." As Professor Wanklyn tells us, the quantity of opium is very small; in fact, as regards any possibility of its becoming "a most beneficial physic," that is altogether out of the question, and practically as barefaced a falsehood as the assertion that Fennings' Fever Curer has for nearly fifty years been curing thousands of cases of diseases which "*no other medicine could cure.*" Of course the reference to an Act of Parliament is another piece of bluff, and so artfully worded as to convey to the class of people who read through every word of the glibberish jargon in patent medicine directions as solemnly as if they were committing to memory some choice piece of Scripture, that the Act of Parliament was passed with special reference to the Fever Cure.

The dose of Fever Curer is put down as a wine-glassful at pretty frequent intervals. Seeing that wine-glasses differ so much in size, it is a good thing that there is so little laudanum contained in the bottle of eight ounces, otherwise "death by misadventure," as juries are apt to bring in the verdict (instead of "manslaughter," the correct term), in cases of fatal poisoning through taking over doses of narcotic quack medicines, would result. As it is, reckoning a properly graduated wine-glass at one-and-a-half ounces, the patient has the gratification of getting through his bottle in about five doses. With regard to the commercial value of the bottleful, we are not disposed in this hot weather to weary either ourselves or our readers by working out a sum in fractions, and we will therefore limit ourselves to the statement that the whole of the drugs in an eight-ounce bottle of this stuff would be dear at a penny.

But, though it is a highly objectional proceeding on the part of patent medicine manufacturers, through greed of gain, to so unscrupulously puff their wares —more energetically and persistently even than that arch-impostor, Autolycus, in Shakespeare's *Winter's Tale*—yet it is many, many times more culpable to impair people's constitutions, and in very numerous instances to jeopardise their lives, by causing the loss of valuable time, during which medical skill would have availed to save the sufferers.

THE END.

SUPPLIED TO HER MAJESTY THE QUEEN.

'𝕵𝖔𝖍𝖆𝖓𝖓𝖎𝖘'-The King of Natural Table Waters-'𝕵𝖔𝖍𝖆𝖓𝖓𝖎𝖘'

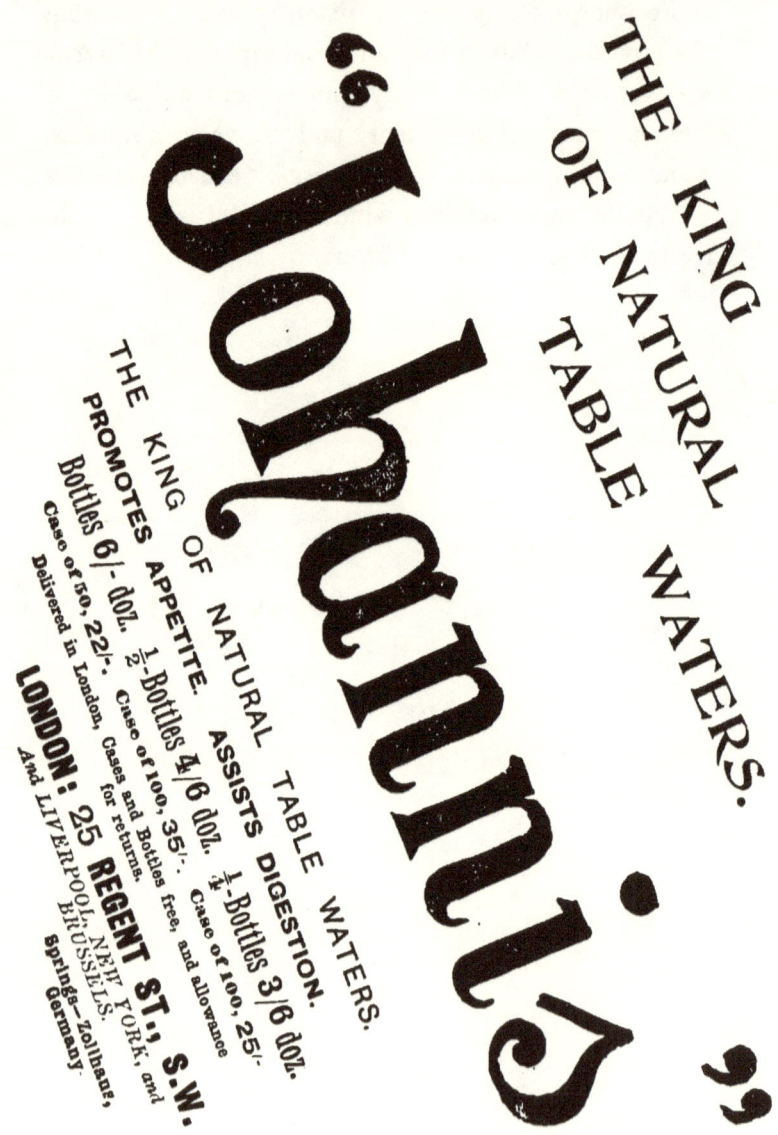

THE
ANTI-ADULTERATION
ASSOCIATION,
(REGISTERED 1895)

61 & 62, CHANCERY LANE, LONDON, W.C.

The frequency and impunity with which the Food and Drugs Adulteration Acts, 1875-72, and other Acts of Parliament having a similar scope, are set at defiance, have led to the formation of this much-needed public Association, the necessity of which has long been recognised, both by the consumers and by the best manufacturers and retailers.

The following are amongst the desirable objects which the Association has in view :—

1.—To put more regularly into force the provisions of the existing Acts, and to prevent their becoming practically inoperative, through the energetic greed of dishonest traders, the apathy or helplessness of consumers, or the inadequate punishment dealt out to offenders in the majority of cases where offences against the Acts have been proved.

2.—To obtain by Statute such modifications of the Acts—at present of too permissive a character, and affording too many loopholes of escape for wrong-doers —as are necessary in the interests alike of the public, and of honest, fair-dealing manufacturers and retailers.

3.—To protect consumers by publishing, from time to time, reports and analyses, together with other useful information concerning foods, beverages, drugs, &c., and their adulteration or purity.

4.—To protect and encourage honest dealers and genuine manufacturers by the issue of official certificates of Purity and Excellence, under certain regulations.

The annual subscription of members is fixed at the small sum of half-a-guinea, entitling the member to HEALTH NEWS, the organ of the Association, sent monthly, post free, and to admission to all lectures and meetings. A donation of £3. 3s., or upwards, constitutes life-membership. Members will also be entitled to have analyses made at reduced terms.

Further particulars, concerning membership, terms for analyses, and regulations for granting certificates can be obtained of the Hon. Secretary. Subscriptions should be made payable by postal order and cheque, crossed "Cheque Bank."

N.B.—Analyses of all kinds conducted in the Laboratory, for non-members, at very reasonable charges.

THE SAVOY PRESS, LTD.,

Savoy House, 115, Strand, W.C.

The following are some of the Publications of this firm:

HEALTH NEWS (Illustrated), Established in 1887. The best, cheapest, and most widely circulated Health Journal. 3d. Monthly, post free for 4 stamps. Subscription for Twelve Months, commencing at any date, 4s., post free to any address in the kingdom.

EXPOSURES OF QUACKERY; containing a Series of interesting Articles upon, and Analyses of, the Principal Patent Quack Medicines. In two volumes, 1s. each, post free for 14 stamps.

STAMMERING, STUTTERING, AND OTHER SPEECH AFFECTIONS. 1s., post free for 14 stamps.

HAY FEVER, HAY ASTHMA, OR SUMMER CATARRH. 1s., post free for 14 stamps.

DEAFNESS, NOISES IN THE EARS, &c. Their Causes and Treatment. 6d., post free for 7 stamps.

CHOLERA: Its Nature, Causes, and Cure. 3d., post free for 4 stamps.

ADVICE TO A WIFE, ABOUT HERSELF AND BABY. 1s., post free for 14 stamps.

FIRST TEN YEARS OF A DOCTOR'S LIFE. 1s., post free for 14 stamps.

GOUT: ITS NERVOUS ORIGIN. 1s., post free for 14 stamps.

INFLUENZA: Its Nature, Symptoms, and Treatment. Second Edition. 1d., post free for 3 Half-penny stamps.

HYDROPHOBIA AND DISTEMPER MADNESS. A Plea for the Canine Race. 1d., post free 3 Half-penny stamps.

DIABETES AND OTHER URINARY AFFECTIONS. 2s. 6d., post free for 2s. 9d.

CADBURY'S COCOA

Absolutely Pure, therefore Best.

GENUINE COCOA.

The public are warned against chemically prepared dark liquor cocoas claiming to be "pure," but in reality prepared with a considerable percentage of alkali; this can be detected by the unpleasant smell when a tin is first opened.

Cadbury's Cocoa, on the other hand, is guaranteed to be absolutely pure, and can be safely and beneficially taken as an article of daily diet AT ALL TIMES AND SEASONS.

THE LANCET SAYS:
"Cadbury's Cocoa represents the Standard of highest purity at present attainable."

www.ingramcontent.com/pod-product-compliance
Lightning Source LLC
Chambersburg PA
CBHW031931230426
43672CB00010B/1889